The Development Dimension

Sustainable Ocean for All

HARNESSING THE BENEFITS OF SUSTAINABLE
OCEAN ECONOMIES FOR DEVELOPING COUNTRIES

This work is published under the responsibility of the Secretary-General of the OECD. The opinions expressed and arguments employed herein do not necessarily reflect the official views of OECD member countries.

This document, as well as any data and map included herein, are without prejudice to the status of or sovereignty over any territory, to the delimitation of international frontiers and boundaries and to the name of any territory, city or area.

The statistical data for Israel are supplied by and under the responsibility of the relevant Israeli authorities. The use of such data by the OECD is without prejudice to the status of the Golan Heights, East Jerusalem and Israeli settlements in the West Bank under the terms of international law.

Note by Turkey
The information in this document with reference to "Cyprus" relates to the southern part of the Island. There is no single authority representing both Turkish and Greek Cypriot people on the Island. Turkey recognises the Turkish Republic of Northern Cyprus (TRNC). Until a lasting and equitable solution is found within the context of the United Nations, Turkey shall preserve its position concerning the "Cyprus issue".

Note by all the European Union Member States of the OECD and the European Union
The Republic of Cyprus is recognised by all members of the United Nations with the exception of Turkey. The information in this document relates to the area under the effective control of the Government of the Republic of Cyprus.

Please cite this publication as:
OECD (2020), *Sustainable Ocean for All: Harnessing the Benefits of Sustainable Ocean Economies for Developing Countries*, The Development Dimension, OECD Publishing, Paris, *https://doi.org/10.1787/bede6513-en*.

ISBN 978-92-64-86891-5 (print)
ISBN 978-92-64-56263-9 (pdf)
ISBN 978-92-64-55639-3 (HTML)
ISBN 978-92-64-79525-9 (epub)

The Development Dimension
ISSN 1990-1380 (print)
ISSN 1990-1372 (online)

Revised version, September 2020
Details of revisions available at: *https://www.oecd.org//about/publishing/Corrigendum.pdf*

Photo credits: Cover © Huy Thoai/Shutterstock.com.

Corrigenda to publications may be found on line at: *www.oecd.org/about/publishing/corrigenda.htm*.
© OECD 2020

The use of this work, whether digital or print, is governed by the Terms and Conditions to be found at *http://www.oecd.org/termsandconditions*.

Preface

The transition to more sustainable ocean economies is urgent. The ocean makes an important contribution to the global economy, providing jobs to millions of people. More than 90% of world trade uses sea routes. It is also the stage for a growing range of new ocean-related economic activities, and the source of constant innovations. Its role however goes much beyond an economic one. It is a key life-support system, central to human well-being and to a healthy planet. The ocean plays a crucial part in the regulation of our climate and weather, and provides critical ecosystem services which are vital to all living creatures.

Nevertheless, pressures on the ocean are mounting. We are witnessing sea-use change, continued over-exploitation of marine resources, intensifying pollution and, above all, climate change. Many ocean-based sectors have expanded with no sufficient consideration for environmental and social sustainability. These developments put at risk the very ecosystems on which economic sectors depend, they erode opportunities for social inclusion and they have important consequences for developing countries.

Ensuring that these countries can harness the full benefits of sustainable ocean economies and a healthier ocean is a global priority. It means helping 82% of the world's population tap into the opportunities of the ocean economy, including more jobs, cleaner energy, improved food security, and enhanced resilience. It is also essential to improve ocean health globally, as developing countries are home to vast ocean resources that the prevailing unsustainable uses will erode, with high socio-economic and environmental costs well beyond these countries. Costs that will be borne by all the economies of our globalised world. Development co-operation has a critical role in supporting developing countries to achieve sustainable ocean economies as well as in promoting a global ocean economy that is governed by institutional arrangements, policies and financing that enables all countries, especially the poorest and most vulnerable ones, to benefit from expanding economic activities in the ocean.

The Sustainable Ocean for All: Harnessing the Benefits for Developing Countries report, which is part of the OECD Sustainable Ocean for All initiative, provides a new comprehensive analysis for policy makers in developing countries and their co-operation partners, to support a transition to sustainable ocean economies in developing countries. It presents OECD data on trends in selected ocean-based industries, and examples of policy instruments, including the latest data on economic tools to promote ocean conservation and its sustainable use. It also contains the first certified estimates of official development assistance (ODA) for sustainable ocean economies, and the estimates of private finance mobilised through these interventions. Building on this, the report provides thorough analysis and tailor-made recommendations for more ambitious and effective policy frameworks, and for targeted and effective development co-operation and financing that help developing countries achieve sustainable ocean economies.

There is no time left for complacency. The COVID-19 crisis is transforming our world, forcing us to change the way we think, the way we work, and the way we produce and consume. Many ocean-based sectors, such as tourism and shipping, have been severely impacted by the crisis, with acutely felt consequences in developing countries. As the recovery plans will shape the direction of global development for decades to come, we need to use them wisely to 'build back bluer'. Investing in ocean-related sectors in a way that fosters environmental, social, as well as economic sustainability is crucial. It will be at the core of a recovery that can put the well-being and resilience of all people, especially the world's most vulnerable, at the centre of a new dawn of sustainable development. We have a unique opportunity to redesign our relation with the oceans. Let's not waste it!

Angel Gurría
OECD Secretary-General

Foreword

The Secretary-General of the Organisation for Economic Co-operation and Development (OECD) launched in early 2019 the Sustainable Ocean for All initiative led by the OECD Development Co-operation Directorate (DCD) to support developing countries harness the benefits of sustainable ocean economies. This initiative offers original evidence and a policy space to contribute to the international ambition for a transition to a global ocean economy that the poorest and most vulnerable countries can benefit from. It contributes to the ambition of the Agenda 2030 on Sustainable Development and the Sustainable Development Goal 14 on the "conservation and sustainable use the oceans, seas and marine resources for sustainable development". With its special attention on small island developing states (SIDS) and least developed countries (LDCs), the report also directly contributes to SDG 14.7, which focuses on increasing the economic benefits from the sustainable use and conservation of marine resources for these countries.

The Sustainable Ocean for All Initiative aims to: (i) enhance the knowledge base and policy options available to developing countries to help them develop evidence-based policies and tools to achieve sustainable ocean economies that bring economic, social and environmental value; (ii) develop new evidence on development co-operation approaches to foster more effective and co-ordinated actions by international development co-operation actors; (iii) align finance to the sustainable ocean economy, both public, private, international and domestic, through adequate domestic policies as well as better aligned development finance and international co-operation efforts; and (iv) increase opportunities for dialogue and mutual learning across countries and within ocean-related communities around the world, including ministries, agencies, academia, foundations, NGOs and the private sector.

This report was developed by the DCD with the OECD Environment Directorate (ENV) and the Science, Technology and Innovation Directorate (STI) to leverage OECD's multi-disciplinary expertise and unique statistical sources in policy areas relevant to the sustainable ocean economy. This report is the first attempt to bring together this expertise in an integrated fashion, with expert inputs provided from across the Organisation (e.g. fisheries, shipbuilding, shipping, tourism). It builds on the OECD Creditor Reporting System database and expertise on development finance and small island developing states; the OECD database on Policy Instruments for the Environment (PINE) and expertise on biodiversity and environmental policy; and the OECD work on measuring the ocean economy and on science and innovation for the ocean.

The report provides original and comprehensive evidence and policy guidance for policy-makers in developing countries and the international development community to address the growing pressures on the ocean and chart a new course for sustainable development through sustainable ocean economies. It presents new evidence on the economic trends of the ocean economy across developing countries, as well as policy options available for fostering more sustainable ocean economies. It also provides the first official estimates of official development assistance for the sustainable ocean economy, new evidence on how development co-operation is supporting sustainability across the spectrum of ocean-based sectors, and how it could help re-orient private finance for sustainable ocean economies.

As part of this work, *Sustainable Ocean Economy Country Diagnostics* were conducted in four countries in 2019-20: Antigua and Barbuda, Cabo Verde, Indonesia, and Kenya. These countries were selected based on the interest expressed by countries themselves and by members of the OECD Development Assistance Committee, as well as on the potential and political leadership on the sustainable ocean economy. Through the Diagnostics additional information was collected through specific questionnaires and interviews to over 50 representatives from across national administrations, the private sector, academia and civil society.

Working with developing countries, as well as with member countries and emerging economies, the OECD aims to contribute to a global transition to sustainable ocean economies that can preserve the ocean, and its many benefits to humanity, for generations to come.

Acknowledgements

This report is a collaboration between the OECD Development Co-operation Directorate (DCD), the Environment Directorate (ENV) and the Science, Technology and Innovation Directorate (STI).

The report was co-ordinated by Piera Tortora (DCD). Together with Claire Jolly (STI) and Katia Karousakis (ENV) they composed the core team for the report. Strategic guidance was provided by Jorge Moreira de Silva, Director of the Development Co-operation Directorate, as the lead Directorate for the report, with further guidance from Haje Schütte and Jens Sedemund.

The core team co-authored the Overview Chapter and Chapter 1. The lead authors for the other chapters are as follows, and have benefitted from inputs and sections from the core team: for Chapter 2 Claire Jolly and James Jolliffe from the OECD STI; for Chapter 3 Will Symes (formerly with the OECD ENV) and Katia Karousakis from the OECD ENV; for Chapters 4 and 5 Piera Tortora with research assistance from Alberto Agnelli, Veronica Villena and Piero Fontolan from DCD.

The report was prepared for publication under the supervision of Henri-Bernard Solignac-Lecomte and Stacey Bradbury, with graphic design by Stephanie Coic and Sara Casadevall Bellés and editing by Susan Sachs. Thanks also go to Meria Greco and Jessica Voorhees for communications support.

The report was informed by the findings from the Sustainable Ocean Economy Country Diagnostics conducted in Antigua and Barbuda, Cabo Verde, Indonesia and Kenya by an OECD cross-directorate team led by Piera Tortora and also composed by Will Symes, James Jolliffe, Piero Fontolan, Alberto Agnelli and Veronica Villena. The team would like to acknowledge the governments of Antigua and Barbuda, Cabo Verde, Indonesia and Kenya for their invaluable collaboration, as well as the following people for the logistical support for the fact-finding missions: High Commissioner Karen-Mae Hill (High Commission for Antigua and Barbuda), Ambassador Helena Paiva and Ms. Helena Guerreiro (Embassy of Portugal in Cabo Verde), Mr. Massimo Geloso Grosso and Ms. Yulianti Susilo (OECD – Jakarta Office), Ambassador Vittorio Sandalli, Mr. Giovanni Brignone and Ms. Mutiara Ronci (Embassy of Italy in Indonesia), Ambassador Alberto Pieri (Embassy of Italy in Kenya), Mr. Hubert Perr (Delegation of the European Union in Kenya). The team would also like to acknowledge the Swedish Ministry of the Environment and Energy for their financial support enabling the Sustainable Ocean Economy Country Diagnostics of Indonesia.

The core team would like to gratefully acknowledge the comments received from OECD colleagues on earlier versions of the report, including: Anthony Cox, Simon Buckle, Edward Perry, Kate Kooka, Andrew Prag, Lisa Danielson, (OECD Environment Directorate), Claire Delpeuch, Haengnok Oh, Nomoto Kazuhiro (OECD Trade and Agriculture Directorate), Andrea Goldstein (OECD Economics Department), Peter Haxton (OECD Centre for Entrepreneurship, SMEs, Regions and Cities), Paul Horrocks, Valentina Bellesi, Juan Casado Asensio, Takayoshi Kato, Heiwon Shin, Carolyn Neunuebel, Özlem Taskin, Tomas Hos, Arnaud Pincet (OECD Development Co-operation Directorate), Pieter Parmentier, Laurent Daniel (OECD Science, Technology and Innovation Directorate).

The authors would also like to gratefully acknowledge comments and feedback from delegates to the Environmental Policy Committee's (EPOC) Working Party on Biodiversity, Water and Ecosystems (WPBWE), delegates to the Development Assistance Committee (DAC), and delegates to the DAC

Network on Environment and Development Co-operation (ENVIRONET). The authors also acknowledge and thank the kind inputs from ministries, departments and research institutes' representatives that form the Steering Board of the STI Ocean Economy Group, overseen by the OECD Committee for Science and Technological Policy (CSTP).

The authors would like to acknowledge the following experts for their comments on the Overview and Chapters 4 and 5: Adrien Vincent (Systemiq), André Rodrigues de Aquino (World Bank Group), Ben Hart (World Resources Institute), Chip Cunliffe (AXA XL), Dennis Fritsch (UNEP FI), Gail Hurley (Global Ethical Finance Initiative), Heino Nau (European Commission DG MARE), Martha McPherson (UCL Institute for Innovation and Public Purpose) and Nicolas Pascal (Blue Finance).

Editorial

Ocean-based sectors, such as fisheries, tourism and maritime transportation, are at the heart of the economy of many developing countries. The way these sectors will develop in the future could either accelerate progress towards sustainable development or exacerbate the current unsustainable trends. Prior to the COVID-19 crisis, the OECD projected that the ocean economy would double in size by 2030. Only if managed sustainably and through recovery efforts that build back bluer, such expansion could lead to more resilient and equitable societies.

Transitioning to sustainable ocean economies is possible, urgent, and delivers universal benefits. The ocean is the largest natural carbon sink on the planet, and a tremendous source of economic livelihoods for billions of people. New solutions for sustainable aquaculture and mariculture can create more jobs and contribute to food security. Renewable ocean energy sources can improve countries' energy mixes. Actions to conserve mangroves and other natural buffers against climate change and extreme weather events can help combat climate change and ensure the delivery of multiple other invaluable marine ecosystem services. For this to happen, international co-operation and global rules and norms on ocean need to be supportive of developing countries' sustainable ocean economies, enhancing their access to ocean science, evidence-based policy advice, innovations and financing to ensure that both existing and emerging ocean-based sectors contribute to inclusive and sustainable development.

Shifting to sustainable ocean economies implies bold and proactive policies that can help refocus economic activities and foster new sustainable economic models that conserve scarce ocean resources and produce more sustainably. But effective policy making for sustainable ocean economies – in both developing countries and the international development community supporting them – is currently hampered by a lack of adequate evidence and a lack of long-term actions across the spectrum of policy areas relating to the ocean. There is a need to develop clear definitions, standards and principles for effective policy making on the sustainable ocean economy, to better track policies and finance, and to share good practices and lessons learned across countries and communities of practice. This report is a step in that direction.[1]

Sustainable ocean economies are far from being a reality in developing countries nor OECD countries. Yet, this report shows numerous positive examples: scientific and technological advances help to map and better understand ocean resources, thus helping to manage them more effectively; the application of economic instruments – including fiscal measures – to encourage a more sustainable use of ocean resources in key economic sectors; development co-operation interventions in support of the deployment of new technologies to enhance sea surveillance and security; measures to increase the value of fish products through certifications of sustainability; and, support to reduce the environmental impacts and GHG emissions from ships and port infrastructure. Much is left to be done, but these examples could be scaled up and provide a basis from which to build a much-needed coherent and effective approach to sustainable ocean management and effective development co-operation for sustainable ocean economies.

The ocean has been the source of wealth and well-being for humans for millennia - we have a responsibility to make sure it will continue to be so forw future generations, benefitting all of humanity.

Rodolfo Lacy

Director,

Environment Directorate

Jorge Moreira da Silva

Director,

Development
Co-operation Directorate

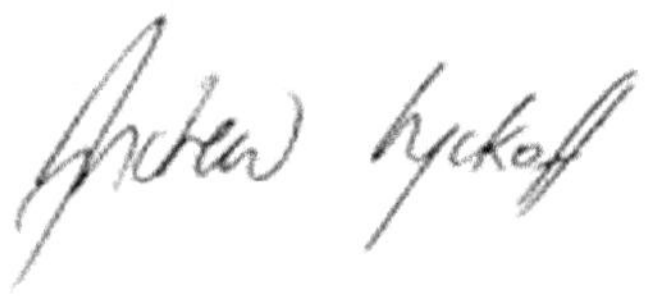

Andrew Wyckoff,

Director,

Science, Technology and
Innovation Directorate

[1] As part of this work, a new data platform on official development assistance for the sustainable ocean economy will be available here: oe.cd/oceandev. An overview of OECD sources on the ocean is available here: https://www.oecd.org/ocean/. Additional OECD sources include OECD ocean data and indicators [OECD (2020), "Sustainable ocean economy", in Environment at a Glance Indicators, OECD Publishing, Paris, https://doi.org/10.1787/1f798474-en.

Table of contents

Preface	3
Foreword	5
Acknowledgements	7
Editorial	9
Executive summary	15
Overview	19
1 The transition to a sustainable ocean economy is a global imperative	**29**
A sustainable ocean economy is a global priority	30
Developing countries stand to gain from a transition to sustainable ocean economies	31
The pressures on ocean and ecosystem services are rising	31
The sustainable ocean economy is multidimensional and rests on three pillars: economic, environmental and social	35
References	39
Annex 1.A. Country groupings	42
2 Developing countries and the ocean economy: Key trends	**45**
Positioning of developing countries in ocean-based industries: Trends across country groupings	46
Trends in selected ocean-based industries	54
Looking forward: Encouraging a balanced mix of sustainable ocean use and conservation	66
References	67
3 Policy instruments and finance for developing countries to promote the conservation and sustainable use of the ocean	**73**
The need for coherent policy approaches to foster sustainable ocean economies	74
An overview of regulatory, economic and other instruments for the conservation and sustainable use of the ocean	75
Policy incentives and finance to conserve the ocean, foster sustainable fisheries, aquaculture and tourism, and manage pollution	81
Science, technology and innovation for sustainable ocean economies	99
References	101
Annex 3.A. Examples of policy instruments to address different pressures on the ocean	109
Notes	110

4 Development co-operation for sustainable ocean economies   111

Development co-operation has a critical role in supporting developing countries to harness the benefits of sustainable ocean economies   112

Growing international attention but no common understanding of the sustainable ocean economy to guide development co-operation   114

Towards common definitions: tracking ODA trends for sustainable ocean economies   116

Mapping the range of sustainable ODA activities for the sustainable ocean economy across key areas   129

Enhancing the impact of development co-operation for sustainable ocean economies   137

References   139

Annex 4.A. Methodology for estimating ocean-relevant ODA   141

Notes   144

5 How development co-operation can help re-orient private finance and investments towards a more sustainable ocean economy   147

The need to increase resources goes hand-in-hand with the need to re-orient private finance away from harmful activities   148

Quantifying private finance mobilised for the sustainable ocean economy through development finance instruments   149

Innovative financial instruments for the sustainable ocean economy   152

Suggestions for re-orienting investments towards sustainability   164

References   167

Notes   170

FIGURES

Figure 1. Growth in ocean-relevant environmental policy instruments   22
Figure 2. ODA to the ocean economy and ODA to the sustainable ocean economy, 2013-18   23
Figure 3. ODA is helping develop new financial instruments and mobilise private finance for sustainable ocean economies   24
Figure 2.1. Global value added in six ocean-based industries, by country income group, 2005-15   48
Figure 2.2. Share of GDP of value added from six ocean-based industries, by country income groups, 2005-15   49
Figure 2.3. Value added of selected ocean-based industries in least developed countries (LDCs)   50
Figure 2.4. Value added from six ocean-based industries, by region and country grouping   51
Figure 2.5. Employment in six ocean-based industries, 2005-15, by region and country grouping   52
Figure 2.6. Share of value added and employment for six ocean-based industries in selected Caribbean countries, 2015   53
Figure 2.7. Value added from six ocean-based industries in selected Caribbean countries   53
Figure 2.8. Global value added in marine fishing by country income group, 2005-15   56
Figure 2.9. Global employment in marine fishing, 2005-15   57
Figure 2.10. Top 20 countries in seafood production in live weight tonnes, 2005-15   58
Figure 2.11. Annual cruise passenger visitors to Caribbean island countries, 2000-18   60
Figure 3.1. Growth in ocean-relevant environmental policy instruments   80
Figure 3.2. Potential structure of a conservation trust fund   84
Figure 3.3. Components of a successful concession system   96
Figure 4.1. Key figures on ODA for the ocean economy and ODA for the sustainable ocean economy (2013-18)   113
Figure 4.2. Key figures on ODA to reduce ocean pollution from land (2013-18)   114
Figure 4.3. Three key indicators to track ocean-relevant ODA   117
Figure 4.4. Ocean-relevant ODA peaked in 2017 but still only accounts for a fraction of global ODA   119
Figure 4.5. SDG 14 is among the least funded SDGs by both Official Development Assistance and philanthropic development funding   120
Figure 4.6. Share of ocean economy ODA that promotes gender equality   121
Figure 4.7. Top providers of ocean economy ODA   123

Figure 4.8. A few development partners provide the bulk of ocean economy ODA 124
Figure 4.9. Top recipients of ocean economy ODA receive small shares of funding integrating sustainability 125
Figure 4.10. ODA is used rather dichotomously, either funding ocean-based industries with no focus on sustainability in large recipients or ocean conservation through small allocations in small recipients 126
Figure 4.11. For most SIDS, ODA for the sustainable ocean economy makes for a small part of the ODA they receive 128
Figure 4.12. Ocean economy ODA to LDCs integrates sustainability by 58% on average 129
Figure 4.13. How development co-operation is helping enhance sustainability across six sustainable ocean economy areas 131
Figure 5.1. ODA mobilised private finance for the sustainable ocean economy largely in upper-middle income countries 150
Figure 5.2. Amounts of private finance mobilised for the ocean through official development assistance 151
Figure 5.3. Key facts about private finance mobilised for the ocean, by ODA instrument 152

INFOGRAPHICS

Infographic 1. Key recommendations to achieve sustainable ocean economies 18

TABLES

Table 1.1. Marine and coastal ecosystem services and the geographic scale of benefits 32
Table 1.2. SDG 14 can contribute to and is strongly linked with other SDGs 36
Table 2.1. Ocean-related economic activities 47
Table 2.2. GDP dependence on selected ocean-based industries 54
Table 2.3. Public perceptions of deep geothermal energy 65
Table 3.1. Policy instruments to promote the conservation and sustainable use of the ocean 76
Table 3.2. Pressures on ocean and marine ecosystems and instruments to address these 76
Table 3.3. Marine spatial planning principles used by the Seychelles 77
Table 3.4. Active ocean-relevant economic instruments in the OECD PINE database 80
Table 3.5. Marine protected area fees in Kenya 82
Table 3.6. Overview of select conservation trust funds 84
Table 3.7. Revenue from access fees for foreign fishing vessels in Pacific Island countries and territories in USD, 2014 90
Table 5.1. Overview of blue bonds and other bonds with blue elements 156
Table 5.2. How development co-operation providers can help replicate and scale up innovative financial mechanisms and scale down unsustainable investments 164

Annex Table 3.A.1. Examples of regulatory policy instruments to address different pressures on the ocean 109
Annex Table 3.A.2. Examples of economic and information and voluntary policy instruments to address different pressures on the ocean 110
Annex Table 4.A.1. Emerging and established sectors and industries of the ocean economy 142
Annex Table 4.A.2. Ocean and ocean economy keywords used for Indicator 1 143
Annex Table 4.A.3. Keywords and filters used for Indicator 2 144
Annex Table 4.A.4. Keywords used for Indicator 3 144

Follow OECD Publications on:

    http://twitter.com/OECD_Pubs

    http://www.facebook.com/OECDPublications

    http://www.linkedin.com/groups/OECD-Publications-4645871

    http://www.youtube.com/oecdilibrary

    http://www.oecd.org/oecddirect/

This book has...   **StatLinks**
A service that delivers Excel® files from the printed page!

Look for the *StatLinks* at the bottom of the tables or graphs in this book. To download the matching Excel® spreadsheet, just type the link into your Internet browser, starting with the *http://dx.doi.org* prefix, or click on the link from the e-book edition.

Executive summary

Achieving sustainable ocean economies has become a global priority. The ocean is at the centre of the livelihoods of more than 3 billion people worldwide, and it is a key life-support system for all life on this planet. However, pressures on the ocean and the ecosystem services it provides have mounted significantly — from overfishing, pollution, and climate change — and are expected to further grow as the ocean becomes the stage for a range of new ocean-related economic activities. These pressures are pushing the health of the ocean to its limits, leading to habitat degradation, ocean warming and acidification, more frequent extreme weather events, and species extinctions. They undermine the ocean's ability to support long-term socio-economic benefits and sustainable development. Actions to revert these trends are therefore urgently needed: new and traditional ocean-based economic activities need to use ocean resources sustainably and conserve them. In recognition of this, for the first time the world agreed to focus on the ocean in the 2030 Agenda for Sustainable Development, through a dedicated Sustainable Development Goal (SDG 14) and ocean action has since become a key priority in international fora, including in recent G7 and G20 agendas.

Achieving sustainable ocean economies goes beyond reaching environmental sustainability. Intrinsically connected with many other SDGs, the conservation and sustainable use of the ocean needs to unlock sustainable development across social, environmental and economic dimensions. It needs to benefit all countries, especially the poorest and the most vulnerable ones, which are highly exposed to the effects of ocean degradation while possessing the least ability to respond. It needs to refocus economic activities and generate new economic and business models that innovatively contribute to both production and ocean health. If pursued in this way, sustainable ocean economies hold the potential to advance progress on key challenges the world faces today: eradicating hunger and extreme poverty, creating more and better jobs, and combatting climate change.

This report brings together OECD's unique statistical sources and expertise across various policy areas to provide a new and comprehensive baseline of evidence to support developing countries and the international development community unlock this potential of the sustainable ocean economy, turning both emerging and existing ocean-based sectors into catalysts for long-term, inclusive, resilient and sustainable development.

The report highlights that the economies of developing countries rely more, on average, on ocean-based sectors for income and jobs than OECD countries. Data collected for this report on six ocean-based industries[1] show that they contribute to more than 11% of GDP in lower middle-income countries and 6% of GDP in low-income countries, compared to less than 2% of GDP for high-income countries in 2015. In some low-income countries and small island developing states, tourism alone and other important ocean-based sectors can account for over 20% of GDP, compared to 2% for OECD countries. Because of this greater reliance on ocean-based sectors, developing countries are likely to face greater risks from rapidly deteriorating marine ecosystems. The evolution of ocean-based sectors could either accelerate progress towards sustainable development or reinforce current trends of environmental degradation and social exclusion.

The report finds significant variations across developing countries, in the potential benefits and risks of a growing global ocean economy. So far, the East Asia and Pacific region recorded the fastest acceleration in value added of the ocean economy, from USD 157 billion in 2010 to over USD 175 billion in 2015, largely driven by the People's Republic of China. In low-income and lower middle-income countries the marine fisheries sector is particularly dominant, accounting for 6% and 8% of GDP respectively. Some countries are home to vast untapped marine resources, but emerging ocean-based sectors, such as deep seabed mining, could have potentially huge negative environmental impacts. It is important that developing countries therefore be in a position to assess and balance the risks and rewards associated with developing these industries and integrate community interests as well as sustainability and environmental concerns from the outset.

Achieving a sustainable ocean economy requires a better alignment of policies across multiple sectors. If policy making across the ocean economy deals with sectors in isolation and without a coherent conceptual frame, multiple and sometimes conflicting policy goals can emerge. Such fragmented policy making will not be sufficient to bring about the urgent and systemic changes required for a sustainable ocean economy. Rather, more holistic and integrated policy approaches are needed to ensure policy coherence, identify and manage trade-offs between sector-specific objectives, and take advantage of synergies where policies can deliver benefits to multiple sectors. The report presents an array of both sector-specific and cross-sectoral policy options to steer developing countries' ocean economies towards sustainability, including marine spatial planning, ecosystem-based fisheries management systems, and economic instruments such as taxes on plastic waste and marine pollution, fees and charges on the use of marine resources, and other incentives for sustainable tourism. The report highlights the importance of factoring in the value of ecosystems into economic decision-making frameworks, scaling up public finance in innovative ways, and investing these resources more efficiently and strategically. Several economic instruments are being used to promote a more sustainable production and consumption patterns across ocean-related sectors, but there is substantial scope to scale these up. These instruments also can generate revenue that could be channelled back into ocean conservation, used to fund sustainable investments directly or to actively co-create markets and tilt the playing field in the direction of sustainability.

However, developing countries often have limited access to the finance, policy evidence, innovations and science needed to realise sustainable ocean economies. Therefore, development co-operation has an essential role to play to facilitate such access, as well as to promote international policy coherence and a reset of international finance for the global ocean economy to truly integrate sustainability and benefit developing countries. Despite the international community's growing interest in ocean matters, the report finds that there remains a lack of common understanding, definitions and principles with which to align co-operation efforts in support of sustainable ocean economies. While some providers have a long track record of support for specific ocean-based sectors, few have a holistic understanding of the sustainable ocean economy, and most lack dedicated development co-operation strategies as well as implementation and monitoring tools.

To help quantify development finance for the sustainable ocean economy and map the range of sustainable activities, this report introduces the first official tracking of official development assistance (ODA) in support of the ocean economy. Furthermore, it provides an estimation of how much of this contributes to a sustainable ocean economy by integrating sustainability. According to this analysis, over the 2013-18 period, an average of USD 3 billion of ODA a year was allocated for the ocean economy and an even smaller volume, USD 1.5 billion on average a year, was allocated in support of the sustainable ocean economy, equivalent to a mere 0.8% of global ODA over the same period. Not only are ODA flows for the sustainable ocean economy small, they could also be employed more effectively. Currently, they are highly concentrated in three sectors – maritime transport, fisheries and marine protection – suggesting that more could be done to support the range of existing and new ocean-based sectors and thus foster greater economic diversification and resilience. Further, increases in ODA for the ocean economy have supported the expansion of ocean-based industries – largely maritime transport –without increasing the sustainability of these sectors. The report identifies the range of actions to enhance sustainability and specific examples across six ocean-based sectors/areas: sustainable fisheries, sustainable tourism, renewable ocean energy, green shipping, ocean conservation and the reduction of ocean pollution. These examples contribute to a common understanding of what constitutes sustainable activities across ocean-based sectors and offer insights into replicable practices.

In addition to directly funding sustainable projects, development co-operation is helping to align private finance to the sustainable ocean economy. Through the catalytic use of guarantees, syndicated loans and other ODA instruments, in 2013-17 development co-operation leveraged USD 2.96 billion of private finance for the ocean economy, 43% of this supporting ocean-based industries and ecosystems and 57% for land-based activities that reduce negative impacts on the ocean such as waste management, sanitation and water treatment. Development co-operation has also proved critical for developing innovative financial instruments for attracting financing for sustainable ocean economies, as blue bonds, debt for ocean swaps, innovative insurance schemes and ocean impact investment funds. Increasing funding towards the sustainable ocean economy needs to go hand in hand with diverting and re-orienting finance away from harmful and unsustainable activities, which often have the greatest negative impacts in developing countries. Therefore, development co-operation needs to support policies, regulations and financial levers to integrate sustainability requirements into traditional financial services and investments, in financial markets (e.g. stocks and bonds), alongside credit markets (e.g. loans or bonds).

Sustainable ocean economies are far from being a reality in developing countries just as in OECD countries, but this report provides a baseline of evidence and numerous positive examples to build a more coherent and effective approach to sustainable ocean management and effective development co-operation in its support to allow developing countries to harness the benefits from sustainable ocean economies.

[1] Marine fishing, marine aquaculture, marine fish processing, shipbuilding, maritime passenger transport and freight transport.

Infographic 1. Key recommendations to achieve sustainable ocean economies

TO HELP DEVELOPING COUNTRIES HARNESS THE BENEFITS OF A MORE SUSTAINABLE OCEAN ECONOMY...

✓ DEVELOPING COUNTRIES CAN...

- Develop holistic, cross-sectoral national sustainable ocean economy strategies and financing strategies
- Promote an integrated approach to management of land, coastal and sea areas
- Get involved with ocean knowledge and innovation networks

✓ DEVELOPMENT CO-OPERATION PROVIDERS CAN...

- Track ODA for the sustainable ocean economy, based on common definitions and principles
- Explore new development schemes fit for achieving sustainable ocean economies
- Use ODA to improve the commercial viability of investments in sustainable activities and businesses
- Integrate ocean sustainability requirements in all ODA and development finance institution (DFIs) lending
- Promote international policy coherence for a sustainable global ocean economy

Overview

The urgent need to transition towards a sustainable ocean economy

Conserving and sustainably using the ocean has gained international momentum. In 2015, global leaders agreed on Sustainable Development Goal (SDG) 14 that specifically targets the ocean, calling for "conservation and sustainable use the oceans, seas and marine resources for sustainable development" (UN, 2015[1]). The G7 as well as bilateral and multilateral initiatives identify ocean action as a key priority.

A healthy ocean is central to human health, well-being and economic activity through the many and invaluable functions and ecosystem services it provides. More than 3 billion people rely on the ocean for their livelihoods worldwide, and an expected global population of 9 billion by 2050 will increase pressure to produce more food, energy and jobs from the ocean. Yet the cumulative impacts of anthropogenic pressures are already pushing the ocean to unprecedented conditions of further warming and acidification, decline in oxygen, and species decline (IPCC, 2019[2]; IPBES, 2019[3]). Pressures are also threatening the ability of marine ecosystems to provide invaluable benefits such as pollution control, storm protection, habitat for species, shoreline stabilisation and flood control. Once-in-a-century extreme sea level events are expected to become once-a-year events in many regions by the middle of this century. These effects are the face of a global ocean crisis that is jeopardising not only the future socio-economic benefits that society derives from the ocean but all life on this planet.

No one knows for how much longer the ocean can continue, under business-as-usual scenarios, to perform its critical functions and provide ecosystem services. The impacts of human activities are long-lasting and, in some cases, irreversible, and further delays in action will result in increasing costs over time. Achieving the conservation and sustainable use of the ocean, seas and marine resources is therefore a matter of urgency. While the level of risk varies across countries, it is likely that all countries, developed and developing, face some level of risk if current trends are not reversed. Urgent actions, therefore, are required of all countries and at multiple levels – local, national, regional and international and from individual citizens – to steer the global ocean economy towards sustainably. If conserved and sustainably used, the ocean can regenerate, be more productive and resilient, and support more equitable societies.

A global transition towards sustainable ocean economies could greatly benefit developing countries, many of which already rely heavily on ocean-based industries and are thus particularly vulnerable to the deterioration of marine habitats and ecosystems on which these industries depend. Sustainable ocean economies should be pursued with a view to achieving poverty reduction, enhancing food security, and achieving decent jobs, clean energy, and healthy and resilient ecosystems and to combat climate change.

This report provides original and comprehensive evidence to support developing countries, and the international development community, to harness the benefits of sustainable ocean economies. The new evidence in the report focuses on:

- significant trends in the ocean economy in developing countries, with an original breakdown by countries' income groupings for six ocean-based industries. Trends in sectors of particular interest to developing countries are also assessed, and include marine fishing and aquaculture, coastal

- and marine tourism, extractive industries (e.g. oil and gas and seabed mining), transport and logistics industries (freight and passenger transport), shipbuilding, renewable energy, and bio-marine resources.
- the need for coherent frameworks and the policy instruments to promote the conservation and sustainable use of the ocean, covering a range of regulatory and economic instruments and other approaches. Emphasis is placed on economic instruments that can provide the correct incentives for sustainable production and consumption and also generate and mobilise finance for the conservation and sustainable use of the ocean.
- the first estimation of Official Development Assistance (ODA) and a mapping and analysis of development co-operation activities for sustainable ocean economies, to promote more effective development co-operation in support of developing countries' transition to sustainable ocean economies and a global ocean economy that benefits developing countries.
- an analysis of how development co-operation is helping mobilise private finance for sustainable ocean economies through grants, guarantees, and other leveraging instruments, and a review of innovative financial instruments, developed with the support of development co-operation, to scale up finance for the ocean. Emphasis is placed on the need for development co-operation to help re-orient financial flows away from destructive practices by supporting the integration of sustainability requirements into traditional financial services and investments, in financial markets (e.g. stocks and bonds), alongside credit markets (e.g. loans or bonds).

The report integrates findings from the Sustainable Ocean Economy Country Diagnostics conducted for this work in Antigua and Barbuda, Cabo Verde, Indonesia, and Kenya. In the course of each of these country diagnostics, additional information was collected, including through interviews with more than 50 representatives from national administrations, the private sector, academia and civil society. The economics trends highlighted in Chapter 2 are based on original OECD long-term experimental time series of value added and employment that were developed for a subset of the ocean economy by the STI Ocean Economy Group and overseen by the OECD Committee for Science and Technological Policy. This subset includes, for now, six ocean-based industries: marine fishing, marine aquaculture, marine fish processing, shipbuilding, maritime passenger transport and maritime freight transport. Many of these industries have strong relevance for developing countries. Chapter 3 builds on unique data available in the OECD Policy Instruments for the Environment (PINE) database that the OECD Environmental Policy Committee oversees and to which more than 110 countries are contributing. Chapters 4 and 5 build on the first quantification and analysis of ocean-related ODA, developed for this report on the basis of the OECD Creditor Reporting System. These two chapters also build on the findings from the OECD Survey on DAC Members' Policies and Practices in Support of the Sustainable Ocean Economy (hereinafter the OECD Survey).

The state of play of sustainable ocean economies in developing countries: Key messages

While the ocean is central to the welfare and prosperity of the whole of humankind, on average developing countries rely on the ocean for jobs and their GDP to a greater extent than do richer countries. OECD countries capture most of the global value added from the ocean economy. In 2015, this amounted to approximately USD 200 billion for a subset of the ocean economy composed of six main ocean-based industries (marine fishing, marine aquaculture, marine fish processing, shipbuilding, maritime passenger transport and freight transport). However, these six industries contribute to a much larger share of developing countries' gross domestic product (GDP): more than 11% of GDP for lower middle-income countries and 6% for low-income countries, compared to less than 2% of GDP for high-income countries in 2015. In some low-income countries and SIDS, other important ocean-based sectors, such as tourism, can account for more than 20% of GDP, compared to 2% for OECD countries.

Across developing countries, the potential benefits and risks from a growing global ocean economy vary largely. Value added of the ocean economy increased the most in the East Asia and Pacific region, growing from USD 157 billion in 2010 to over USD 175 billion in 2015 and largely driven by the People's Republic of China (hereafter China). The sectoral composition of the ocean economy also differs across countries, with marine fisheries particularly dominant in low-income and lower middle-income countries where this sector represents up to 6% and 8%, respectively, of GDP. Beyond the six selected industries, two additional ocean-based industries – tourism and offshore oil and gas – are crucial for low-income countries and small island developing states (SIDS). Two out of three SIDS rely on tourism for 20% or more of GDP. The total contribution of tourism to GDP exceeds 40% in Antigua and Barbuda, Belize, Maldives, Saint Lucia, and Fiji and more than 50% of GDP in Cabo Verde and approximately 65% of GDP in the Seychelles. Offshore oil production and its supporting activities are also crucial for many African, Latin American and South Asian countries. In Angola, it contributes about 50% of the country's GDP and about 89% of exports. Looking forward, some emerging ocean-based sectors, such as deep seabed mining, could have potentially huge environmental impacts. Developing countries must be in a position to assess and balance the risks and rewards associated with developing these industries to achieve a sustainable use of resources that effectively integrates community interests and environmental concerns from the outset.

The strong reliance of some developing countries on the ocean makes them particularly vulnerable to growing anthropogenic pressures on the ocean and increases the need to pursue sustainable development models. Many developing countries are home to vast untapped reserves of ocean-based minerals, natural gas and seafood as well as to marine biodiversity hotspots. Shifting towards more sustainable use and conservation of ocean resources is essential, not only to ensure ocean health globally, but also to allow developing countries to fully benefit from the opportunities of the ocean economy. Conversely, using ocean resources through the prevailing harmful and unsustainable practices would lead to high socio-economic and environmental costs for developing countries and the world's ocean. If developing countries are to be able to harness the benefits of the sustainable ocean economy, economic activities (such as ocean-based industries, artisanal activities and other land-based industries that impact the ocean) must be steered towards sustainability through adequate policies and regulations at national, regional and international levels.

Developing countries can use a range of policy instruments (regulatory, economic, and information and voluntary approaches) to harness the benefits of sustainable ocean economies. Robust and coherent policy frameworks are key to ensuring the conservation and sustainable use of the ocean. Sectors in the ocean economy are highly interconnected, and actions taken in one sector have impacts on others. If policy making across the ocean economy deals with sectors in isolation and without a coherent conceptual frame, multiple and sometimes conflicting policy goals can emerge. Such fragmented policy making will not be sufficient to implement the urgent and systemic changes required to steer the ocean economy towards greater sustainability. Therefore, more holistic and integrated policy approaches are needed to ensure policy coherence, identify and manage trade-offs between the sectors, and take advantage of synergies where policies can deliver benefits to multiple sectors.

Promoting the ocean economy sustainably requires not only well-designed policies but also adequate finance to ensure these are effectively managed, monitored and enforced. Policies to address the multiple anthropogenic pressures on the oceans include instruments that are cross-cutting, such as marine spatial planning and environmental impact assessments, and that are sector-specific, such as taxes on plastic waste and marine pollution, ecosystem-based fisheries management systems, fees and charges on the use of marine resources, incentives for sustainable tourism, and payment schemes including blue carbon payments for ecosystem services. Economic instruments in ocean-based industries can also generate government revenue that can be channelled towards conservation and sustainable use of the ocean. Globally, more and more countries have introduced economic instruments targeted at ocean sustainability. By 2020, 57 countries had introduced ocean-related instruments, more than three times as many as in 1980, according to data reported to the OECD PINE database (Figure 1). In 2018, at least USD 4 billion was raised through ocean sustainability-related taxes, including on fisheries, maritime transportation and marine pollution.

Figure 1. Growth in ocean-relevant environmental policy instruments

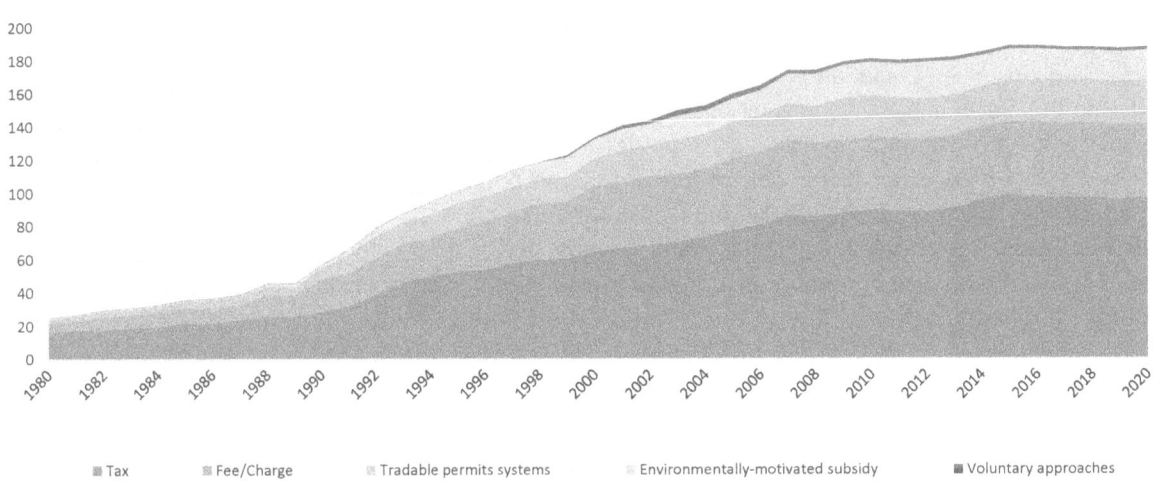

Source: (OECD, 2020[4]), *Policy Instruments for the Environment database*, oe.cd/pine, accessed June 23, 2020.

StatLink https://doi.org/10.1787/888934159373

Scientific and technological advances can contribute greatly to mapping and better understanding ocean resources and thus to developing effective policy frameworks and managing these resources more effectively. Scientific and technological advances are contributing to an integrated and evidence-based management of the ocean and fostering solutions to improve the sustainability of ocean-based economic activities (OECD, 2019[5]). For developing countries, such advances can contribute greatly to building national ocean strategies by enhancing understanding of national marine resources and enabling more effective monitoring of the activities taking place in national waters and beyond.

Developing countries can tap into these advances for the development of sustainable ocean strategies with support from donor countries and marine research institutes around the world. Ongoing scientific exploration campaigns (mapping the seafloor and flora and fauna) are already contributing to improved understanding of the national marine resources of Cabo Verde, Gabon and the Seychelles, for example. Marine technology transfers are also helping countries more effectively monitor the activities taking place in their national waters and beyond, for instance by tracking fishing vessels and shipping. An important first step to assess fragile ecosystems may be linking with existing knowledge and innovation networks, to form partnerships where mutual learning is key, and base any future activities on scientific evidence.

Development co-operation has a unique role to play to ensure that developing countries have access to the knowledge, science, innovations and finance needed to achieve sustainable ocean economies and that the global ocean economy is guided by institutional arrangements and policies aligned to sustainability and benefitting the world's poor and most vulnerable. The development community is increasingly engaged on ocean issues and many bilateral and multilateral initiatives have emerged in the past three years. However, there is a need to develop common definitions, standards and principles to monitor the progress of the development co-operation community in supporting sustainable ocean economies and to promote mutual learning and the adoption of good practices for more effective development co-operation in this area. This is necessary also because while some development co-operation providers have a track record of support for fisheries or other ocean-based sectors, the sustainable ocean economy concept is fairly new, and few providers understand the ocean economy holistically and have dedicated strategies and management tools.

The volume of ODA for the ocean economy is rising, outpacing in 2013-18 the growth rate of global ODA, but it is still small – USD 3 billion on average a year in 2013-18 – and only about half of it promotes sustainability (i.e. ODA for the sustainable ocean economy) (Figure 2). ODA for the sustainable ocean economy is concentrated on three sectors - maritime transport, fisheries and marine protection – rather than supporting a wider range of existing and emerging ocean-based sectors to foster economic diversification and resilience. The report focuses on six areas/sectors of the sustainable ocean economy (i.e. ocean conservation, sustainable fisheries, sustainable tourism, renewable marine energy, and curbing ocean pollution) to illustrate the range of sustainable ODA projects in each of these areas and contribute to a common understanding of what contributes to sustainability with regards to the ocean economy. These sustainable ODA projects, which include new technologies for sea surveillance and security; value addition of fish products through sustainability certifications, support to reduce greenhouse gas emissions from ships and port infrastructure, etc. – also offer insights into replicable and scalable approaches.

Figure 2. ODA to the ocean economy and ODA to the sustainable ocean economy, 2013-18

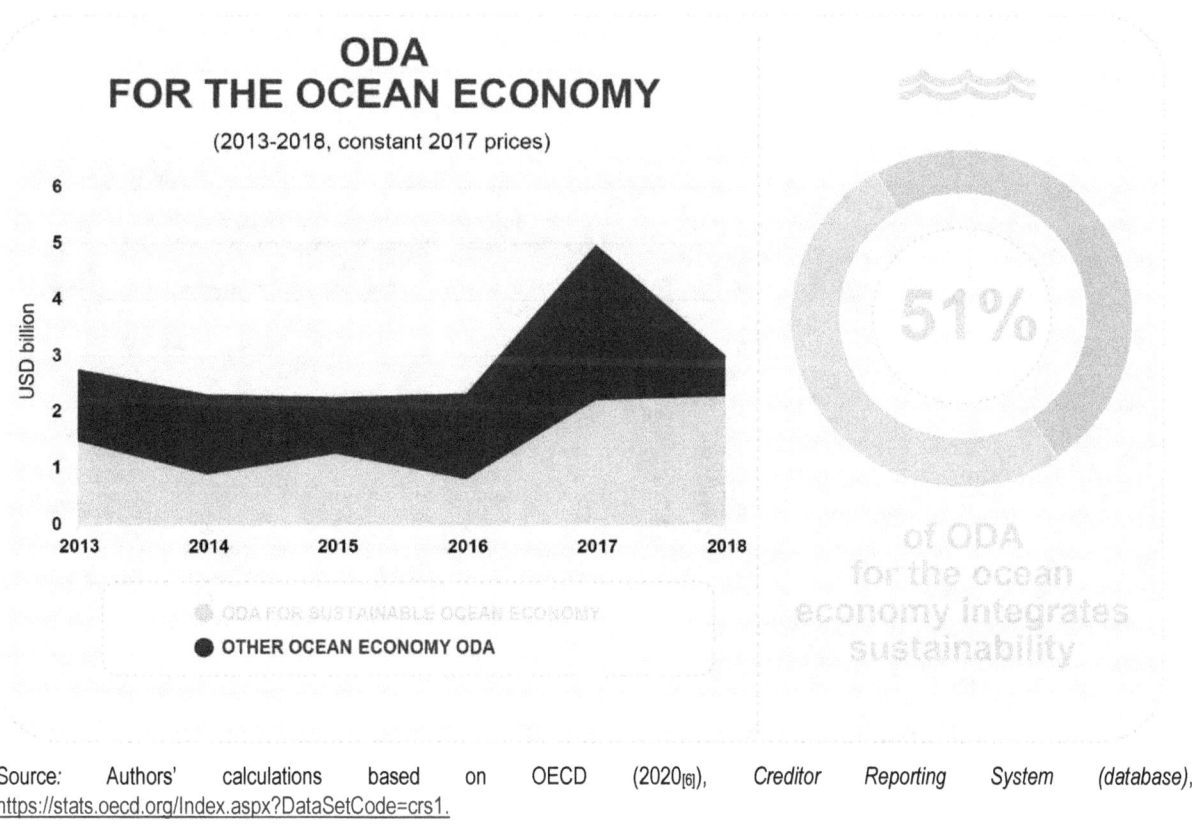

Source: Authors' calculations based on OECD (2020[6]), *Creditor Reporting System (database)*, https://stats.oecd.org/Index.aspx?DataSetCode=crs1.

StatLink https://doi.org/10.1787/888934159107

Development co-operation is leveraging private finance and supporting the development of innovative financial instruments that mobilise capital for the sustainable ocean economy. Through the catalytic use of ODA grants, guarantees and other financial instruments, development partners leveraged a total of USD 2.96 billion of private finance in support of the ocean economy in 2013-17, equivalent to an annual average of USD 593 million. This includes private finance mobilised for ocean-based industries and ecosystems (USD 1.26 billion, or 43%) as well as private finance mobilised for land-based activities that reduce negative impacts on the ocean such as waste management, sanitation and water treatment (USD 1.66 billion, or 57%). (Figure 3 left panel).They also contributed to develop new financing instruments – such as blue bonds, debt-for-nature swaps and new insurance schemes for the ocean Figure 3, right panel)– by providing technical assistance, absorbing the development phase costs of new instruments, using concessional finance to achieve credit enhancement, and supporting the identification of a pipeline of bankable projects. However, greater efforts are need to re-orient financial flows that perpetuate destructive practices that often negatively impact developing countries' fish populations, coasts, tourism, food security and livelihoods. Development partners have a critical role to play to support policies, regulations and financial levers to divert finance from such harmful and unsustainable practices and to integrate sustainability requirements in traditional financial services and investments, financial markets (e.g. stocks and bonds), and credit markets (e.g. loans or bonds).

Figure 3. ODA is helping develop new financial instruments and mobilise private finance for sustainable ocean economies

BETWEEN 2013 AND 2017, ODA MOBILISED USD 2.92 BILLION OF PRIVATE FINANCE IN SUPPORT OF THE OCEAN

LAND BASED 57%
OCEAN ECONOMY 43%
LAND BASED USD 1.66 BILLION
OCEAN ECONOMY USD 1.26 BILLION

DEVELOPMENT CO-OPERATION HAS SUPPORTED THE DEVELOPMENT OF INNOVATIVE FINANCIAL INSTRUMENTS FOR THE SUSTAINABLE OCEAN ECONOMY

BLUE BONDS

The first **Blue Bond** was issued by the Government of Seychelles (USD 15 million) supported by aWB guarantee and GEF loan, to foster the sustainable use of marine protected areas and priority fisheries; expand seafood value chains.

DEBT FOR OCEAN SWAPS

Sovreign debt restructuring for marine conservation in Seychelles (USD 21.6 million). Capital mix: official creditors, TNC, philanthropic foundations.

RISK MANAGEMENT TOOLS

Parametric Insurance for Coral Reefs in Mexico: Insurance policy to protect coral reefs and beaches around Cancun and Puerto Moreios from high-speed winds, developed with support from development partners.

IMPACT INVESTMENT FUNDS

Sustainable Ocean Fund: Impact investment fund (USD 37.5 million) managed by Althella, for funding investments on fisheries, sustainable aquaculture and seafood supply chains.

Meloy Fund: Impact investment fund that conducts debt and equity investments in small fisheries in the Philippines.

Source: Authors

The COVID-19 crisis is severely affecting key ocean-based sectors and casting growing uncertainties on the global ocean economy, but it needs to be turned into an opportunity to rebuild ocean industries more sustainably through co-ordinated efforts and adequate international support. The impacts of the COVID-19 pandemic could have long-lasting effects on ocean-based industries. The acute disruptions to tourism, marine transport and other ocean-based sectors have significant economic ramifications for many developing countries, including some of the most vulnerable countries such as small island developing states (SIDS). While the pandemic is casting a pall of uncertainty over the global ocean economy, it will be important for countries to not lose sight of the longer-term opportunities that a sustainable ocean economy can offer and to consider how to best integrate sustainability into stimulus packages and recovery efforts in national goals and objectives, with support from the international community.

To achieve sustainable ocean economies that benefit all, actions are required of all countries and stakeholders. The following are recommendations adressed to developing couries and development co-operation providers for developing countries to harness the benefits of the sustainable ocean economy.

Developing countries

1. **Identify and monitor key pressures on the ocean under national jurisdiction, where they originate (local, regional, international) and from which sectors:**
   - Identify key pressures on the ocean and the sectors and activities they stem from. When possible, consider both current as well as projected pressures under business-as-usual scenarios.
   - Monitor the state of the ocean including fish stocks and other resources and the condition of coral reef and other marine habitats. Where possible, assess the potential economic opportunities within existing and emerging sectors.

2. **Promote policy coherence via effective national institutional arrangements, including inter-ministerial or inter-agency co-operation and co-ordination, and promote an integrated approach to management of land, coastal and sea areas:**
   - Review and evaluate existing institutions as well as any mechanisms in place to foster national policy coherence, with the aim of aligning the economic, social and environmental objectives.

3. **Evaluate the adequacy of existing policy instruments to address the pressures on the ocean, identify policy gaps and estimate the financing needs to put in place effective policy mixes:**
   - Examine and evaluate the policy instruments that are already in place and identify the policy gaps and opportunities to scale up existing policy instruments. The suite of policy instruments ranges from regulatory instruments to economic, information and other instruments. Economic instruments include fiscal measures such as taxes and subsidies, as well as instruments such as payments for ecosystem services and user fees (e.g. entrance fees for marine protected areas).

4. **Identify the level of finance needed to scale up existing measures and put in place new measures:**
   - Evaluate the role that fiscal and other economic instruments, other finance mechanisms and development finance can play to help to close the finance gap.

5. **Develop national strategies and plans to ensure the conservation and sustainable use of the ocean, including elements to mobilise finance and to monitor progress over time:**
   - Develop national strategies and plans that set a long-term vision for the sustainable ocean economy and recognise the complexity of intersectoral interactions, integrate environmental, social and economic values, and include adequate resources across sectors. In these strategies and plans, establish clear targets, timetables and indicators to monitor progress.

- Track finance generated through fiscal and other economic instruments and through other finance mechanisms. Track their impacts to promote a common understanding of how these instruments and mechanisms can support a sustainable ocean economy.
- Monitor the implementation of policies and their efficacy over time.

6. **Seize the opportunities offered by scientific and technological advances to foster a sustainable ocean economy.**

- Endeavour to get involved with ocean knowledge and innovation networks that are developing around the world, and join international co-operation initiatives to benefit from knowledge transfers and collaboration in capacity building more generally. This calls also for support from developed countries to encourage and facilitate the engagement of low-income countries in such networks and projects. As an example, the Commonwealth Blue Charter promotes ocean exploration and marine mapping projects in interested coastal countries and small island developing states that are conducted in partnerships with United Kingdom oceanographic institutions. On a global basis, the United Nations Decade of Ocean Science for Sustainable Development, to be launched in early 2021, will provide an important opportunity to engage in new research and capacity building programmes.

Development co-operation providers

1. **Develop coherent, evidence-based approaches and tools to support developing countries to manage the risks and opportunities of the ocean economy and transition to sustainable ocean economies:**

- Support developing countries develop a coherent, unified policy strategy that sets out a vision and direction for the sustainable ocean economy and in which the complexity of intersectoral interactions is understood; environmental, social and economic values are integrated; and adequate resources are mobilised across sectors.
- In existing sectors, development co-operation could focus support on correcting the trends of financial leakages, economic exclusion, and environmental degradation in existing ocean-based sectors. In emerging sectors, providers can support countries to develop capacities to assess and balance risks and rewards of new ocean-based sectors so as to achieve a sustainable use of resources that effectively integrates community interests and environmental concerns from the outset.
- Track ODA for the sustainable ocean economy and its impacts based on a common definition and clear principles. The tracking of ocean-relevant ODA should become an integral part of the regular monitoring of ODA flows to provide transparency and accountability of these flows and promote mutual learning on the most effective ODA interventions and approaches for supporting sustainable ocean economies. An official taxonomy of ODA for the sustainable ocean economy could be developed to guide the tracking and monitoring of ODA flows. Ideally such taxonomy would also include gender inclusion and social sustainability criteria.
- Explore new development co-operation schemes fit for transitioning to sustainable ocean economies, for example new co-operation schemes that take into account the global public good nature of ocean resources and development co-operation schemes to strengthen developing countries' expertise and negotiating skills in relation to new market opportunities.

2. **Help align private finance to the sustainability imperative of ocean economies, supporting the scaling up of sustainable investments and helping to redirect private finance away from harmful practices and towards sustainable activities:**

- Use ODA catalytically to improve the commercial viability of investments in sustainable activities and businesses, helping to create new sustainable products and markets including through new investment vehicles and instruments.

- Integrate ocean sustainability requirements in all ODA lending and all development finance institution (DFIs) lending (not all of which is concessional in ODA terms).
- Support the adoption of the Sustainable Blue Economy Finance Principles and the integration of ocean sustainability requirements by international finance institutions (IFIs), which bilateral development partners can influence as they are members and shareholders of these institutions.
- Advocate for the adoption of the integration of ocean sustainability requirements by exchange listing and other financial market regulations for refocusing investments to ocean-based industries towards sustainability and explore additional financial and policy levers.
- Strengthen independent assessments of the impacts of financial flows to the ocean economy such as through international and research institutions.

3. **Promote international policy coherence for a sustainable global ocean economy:**
- Support independent studies on how the sustainable ocean economies of developing countries are affected by OECD countries' policies beyond development co-operation, such as fisheries, tourism, investment and finance policies, as well as by transboundary policies and impacts, such as transboundary pollution.
- Promote evidence-based dialogue across countries on the impact of policies beyond development co-operation on the sustainability of developing countries' ocean economies.

1 The transition to a sustainable ocean economy is a global imperative

Transitioning to sustainable ocean economies is a matter of urgency at the international level and for developing countries. This chapter explores the reasons. It describes in detail the key pressures on the ocean to shed light on why business-as-usual scenarios for the ocean economy are not sustainable and jeopardise socio-economic benefits that can be derived from ocean-based industries and activities as well as life as we know it on this planet. It discusses the challenge of defining sustainable ocean economies, highlighting the need to integrate economic, environmental and social dimensions of sustainability, and the inter-linkages across Sustainable Development Goals that can help reinforce the sustainable use and conservation of the ocean

A sustainable ocean economy is a global priority

The 2030 Agenda for Sustainable Development and the 17 Sustainable Development Goals (SDGs), adopted in September 2015 by the United Nations (UN) General Assembly, put the ocean high on the international political agenda. For the first time, a global development goal specifically targets the ocean with an explicit aim to balance the use and conservation of the ocean use and conservation: SDG 14 calls for the "conservation and sustainable use the oceans, seas and marine resources for sustainable development" (UN, 2015[1]). Also for the first time, a UN Special Envoy for the Ocean was appointed, and in 2017, Sweden and Fiji co-hosted the first UN Ocean Conference, which placed ocean action at the centre of the efforts required to deliver on our people, our planet and our prosperity (UN, 2017[2]). The conference concluded with a 14-point Call for Action that garnered more than 1 400 voluntary commitments towards ocean conservation and sustainable use and reaffirmed the commitment to mobilise resources in line with the Addis Ababa Action Agenda (UN, 2017[2]). More recently, the G7 and G20 have also placed ocean issues on their agendas.

The ocean produces half the Earth's oxygen and absorbs more than 90% of the heat resulting from anthropogenic greenhouse gas emissions, thereby regulating the climate. The ocean also provides habitat for marine species, including many that people depend on for food, and nutrient cycling. It offers hazard protection from natural disasters via shoreline stabilisation, and pollution and flood control. The ocean is also central to the identity, the culture and the economic livelihoods of billions of people.

The ocean is vital to the world's economy, with more than 90% of trade using sea routes, and is a source of jobs for millions of people (International Chamber of Shipping, 2020[3]). It is also the stage for a growing range of new, ocean-related economic activities. Prior to the COVID-19 pandemic, OECD estimates suggested the value added generated by ocean-based industries globally could double from USD 1.5 trillion in 2010 to USD 3 trillion in 2030 (OECD, 2016[4]). In particular, marine and coastal tourism, capture fisheries, marine aquaculture, marine fish processing, and offshore wind and port activities are projected to grow most rapidly in terms of value added globally.

While economic activities continued to accelerate until late 2019, the COVID-19 pandemic has had a halting effect, with large impacts on several countries around the world. Measures meant to control the spread of the disease are affecting ocean activities, not least those at the heart of the global trade and transportation system, and the profound economic effects engendered by the pandemic are likely to continue after the health emergency comes to an end. The precise impacts of this disruption on the future of the ocean economy and on the marine environment are, as yet, unclear. However, economic activity in key ocean-based industries is broadly expected to slow down and it may take several months before pre-crisis levels are reached again.

The COVID-19 pandemic may generate considerable short-term costs, but it also provides an opportunity to rethink and steer economic activities in the ocean towards greater sustainability. Stimulus packages will be decisive in shaping the nature of the recovery and will need to provide a solid basis for relaunching development on a more sustainable footing. Despite the current slowdown in economic activity, demands on marine resources for food, energy, minerals, leisure and other needs of a growing global population will persist. Improving long-term sustainability is thus more urgent than ever before, and it should become a critical factor in decision making surrounding ocean-based industries as policy makers are considering strategies that will stimulate recovery. The recovery and future development of ocean-based industries in the ocean economy should go hand in hand with ensuring that the value of marine ecosystems is maintained for current and future generations.

Developing countries stand to gain from a transition to sustainable ocean economies

Many developing countries[1] rely on ocean-based industries such as tourism and fisheries for foreign exchange, income and jobs. If managed sustainably, the global expansion of existing and new ocean-based sectors until 2019 – which may resume after the COVID-19 pandemic – could potentially advance sustainable development more broadly by creating new opportunities for jobs, food security and clean energy and to achieve diversified and resilient economies. However, this expansion of the ocean economy carries real risks. If greater efforts are not made to curb and re-orient currently destructive economic activities that often have the largest impacts on developing countries' fish populations, coasts and tourism, food security, and livelihoods, the ocean economy could deepen socio-economic injustice and environmental degradation.

Ocean-based sectors have often expanded without sufficient consideration for social and environmental sustainability, creating low-wage jobs and negatively impacting the environment in developing countries. Many developing countries are also especially vulnerable to the impacts of climate change and ocean pollution that further degrade the marine ecosystems on which they rely and often disproportionately affect the poorest segments of their populations. Intricately connected and mutually-reinforcing ocean pressures – ocean pollution including from marine litter, overfishing, warming of the ocean, acidification and loss of oxygen, sea-level rise and more frequent extreme weather events – are already having disruptive effects on coastal communities and entire sectors of developing countries' economies. They also are eroding social inclusion and putting at risk both the resources on which these sectors depend and their future socio-economic benefits.

The impacts on developing countries are part of a global set of impacts that are pushing the ocean, the world's shared life support system, to conditions never before experienced in human history. Tackling the ocean crisis requires urgent actions from all countries and at multiple levels, subnational as well as regional and international, and from individual citizens. However, while developing countries may often suffer the greatest consequences from unsustainable domestic and global ocean-based activities, they often have the least capacity to respond and may not be in a position to enact the comprehensive and urgent actions needed to transition to sustainable ocean economies.

The international development community can play a critical role to ensure that the development of the global ocean economy benefits all, and specifically developing countries. A transition to sustainable ocean economies in developing countries can achieve dual objectives: harnessing the benefits of an expanding global ocean economy to realise sustainable development and, given the cross-boundary nature of ocean challenges, contributing to set the global use of ocean resources on a more sustainable footing. Therefore, development cooperation can help ensure that the development of the global ocean economy is guided by institutional arrangements, policies and financial flows that are aligned with the imperative of sustainability and the needs of developing countries. Development co-operation also can support developing countries turn both emerging and existing ocean-based sectors into catalysts for long-term, inclusive and sustainable development by facilitating access to the science, knowledge, innovations and finance they need to accomplish this endeavour.

The pressures on ocean and ecosystem services are rising

The conservation and sustainable use of the ocean is critically important. Ocean and marine ecosystems provide the intermediate inputs – ecosystem services – that drive value within sectors by acting as nurseries for fish, providing areas for recreation such as beaches and coral reefs for diving, and providing genetic material for marine biotechnology (OECD, 2017[5]) (OECD, 2016[4]). Table 1.1 summarises the ecosystem services provided by the ocean and where their benefits are experienced.

Table 1.1. Marine and coastal ecosystem services and the geographic scale of benefits

Examples of ecosystem services	Geographical scale of benefits		
	Local	Regional	Global
Food provisioning (e.g. fisheries and aquaculture)	✓	✓	✓
Fuel (e.g. mangrove wood)	✓	✓	✓
Natural products (e.g. sand, pearls and diatomaceous earth)	✓	✓	✓
Genetic and pharmaceutical products	✓	✓	✓
Lifecycle maintenance, habitat and gene pool protection			✓
Atmospheric composition, carbon sequestration and climate regulation	✓	✓	✓
Shoreline stabilisation/erosion control	✓	✓	
Natural hazard protection (e.g. from storms, hurricanes and floods)	✓	✓	
Pollution buffering and water quality	✓	✓	
Soil, sediment and sand formation and composition	✓	✓	
Recreation and tourism	✓	✓	✓
Cultural and spiritual values	✓	✓	✓
Scenic beauty	✓	✓	

Source: OECD (2017[5]) *Marine Protected Areas: Economics, Management and Effective Policy Mixes*, https://doi.org/10.1787/9789264276208-en.

Despite the invaluable benefits provided by a healthy ocean, the cumulative impacts of anthropogenic pressures are pushing the ocean to conditions outside human experience. These include warming and acidification and decline in oxygen as well as species decline (IPCC, 2019[6]; IPBES, 2019[7]). Sea levels continue to rise at an increasing rate. Extreme sea-level events of the kind that have historically been rare now are projected to occur more frequently. It is not known for how much longer the ocean can continue to provide its life-sustaining functions and ecosystem services under business-as-usual scenarios.

The state of marine ecosystems was raising alarms a decade ago, when it was estimated that 60% of the world's major marine ecosystems had been degraded or were being used unsustainably (UNEP, 2011[8]). The Intergovernmental Science-Policy Platform on Biodiversity and Ecosystem Service (IPBES) more recently reported that 66% of the ocean area is experiencing increasing cumulative impacts and over 85% of wetlands (area) have been lost (IPBES, 2019[7]). Since the 1980s, for example, an estimated 20% of global mangroves have been lost. More than 30% of global fish stocks are overexploited, and coral reefs are bleaching due to exposure to high temperatures and other pressures. Live coral cover on reefs has nearly halved in the past 150 years, with the decline dramatically accelerating over the past 20-30 years due to increased water temperature and ocean acidification interacting with and further exacerbating other drivers of loss (IPBES, 2019[7]). Moreover, even if global temperatures were to stay within the 1.5°C scenario, it is projected that 90% of global coral would be lost. Concurrently, pollution from land-based sources including marine litter is threatening species and marine habitats. Climate change compounds these effects, altering both the thermal and chemical characteristics of the ocean as well as its dynamics and nutrient availability (Bijma et al., 2013[9]). The welfare costs that this imposes on society are high, and estimates suggest that the cumulative economic impact of poor ocean management practices is on the order of USD 200 billion per year (UNDP, 2012[10]).

Additionally, these pressures can reinforce each other, exerting greater cumulative impacts on marine ecosystems. Global sea-level rise, warming of the ocean, more frequent and severe weather events, and changing ocean currents will aggravate the negative impacts of overfishing; illegal, unreported and unregulated fishing (IUU); pollution; and habitat degradation. In marine ecosystems, according to the IPBES (2019[7]), direct exploitation of organisms (mainly fishing) has had the largest relative impact, followed by land-/sea-use change, including coastal development for aquaculture and infrastructure, and pollution. The remainder of this section outlines the key pressures on the oceans.

Climate change

Anthropogenic carbon dioxide emissions have risen over time, and the ocean has absorbed 20-30% of the carbon dioxide, leading to ocean acidification. Greenhouse gases in the atmosphere have led to rising sea temperatures and sea levels and shifts in ocean currents (IPCC, 2019[6]). There is widespread and mounting concern about the future impact of climate change on the health of the ocean. The implications for ocean ecosystems and marine diversity are considerable, already being seen in species and habitat loss, changes in fish stock composition and migration patterns, and higher frequency of severe ocean weather events. The higher frequency of severe ocean weather events particularly affects vulnerable, low-lying coastal communities, including small island developing states. Nearly 2.4 billion people, or about 40% of the world's population, live within 100 km of a coast and more than 600 million people, or about 10% of the world's population, live in coastal areas that are less than 10 metres above sea level. These populations depend on and are vulnerable to the ocean's quality, stability and accessibility. Developing countries could be much harder hit than industrialised nations by the impacts of climate change. The consequences of climate change on ocean-based economic activities in these countries are being – and will continue to be – felt extensively. These activities include fishing and aquaculture operations, the offshore oil and gas industry, shipping companies, coastal and marine tourism, and marine bioprospecting for medical and industrial purposes.

Pollution

Marine pollution occurs when harmful or potentially harmful impacts result from the entry into the ocean of chemicals; particles; industrial, agricultural and residential waste; noise; or the spread of invasive organisms. Most sources of marine pollution (80%) are land-based (GOC, 2014[11]) and include industrial, residential and agricultural runoffs and waste such as plastics as well as solid waste. In particular, the runoff of agriculture fertilisers, animal husbandry waste, sewage disposal and industrial effluents releases excessive nutrients into the ocean that favour the growth of toxic and harmful species in the ocean (eutrophication), altering marine habitats and negatively impacting fisheries (GESAMP, 2001[12]). Marine pollution also originates from direct discharge through ship pollution (e.g. ballast water and hot water discharge) and deep-sea mining (e.g. for oil and gas), with the resulting types of pollution consisting of acidification, eutrophication, marine litter, toxins and underwater noise. Left unchecked, eutrophication can lead to the creation of dead zones, as is occurring in different parts of the world including the Gulf of Mexico, the Black Sea and the Baltic Sea.

Marine litter, including plastics, is generated directly or indirectly by very different economic sectors, for example aquaculture and fisheries (e.g. accidental loss, intentional abandonment and discarding of fishing gear), shipping and cruise ships (e.g. ship-generated waste), cosmetics and personal care products, textiles and clothing, retail, and increasingly tourism. Illicit dumping particularly affects artisanal fisheries and the tourism industry, as the health and safety of persons who use beaches for recreational activities are at risk in areas where litter accumulates, with both sectors often representing the primary form of foreign revenue for many developing countries.

Plastics are a significant source of ocean pollution: between 4.8 and 12.7 million tonnes of plastics enter the ocean each year (Diez et al., 2019[13]). The cost of ocean plastic is estimated at USD 13 billion per year due to its negative impacts on the coastal environment, tourism and the fisheries industry (OECD, 2018[14]). Despite the growing body of evidence on the negative impacts of marine plastics, there is still considerable uncertainty about the long-term impacts, as large volumes of plastics have only recently been introduced into marine ecosystems. Many of these plastics are extremely long-lived and will remain in the environment for hundreds, if not thousands, of years, meaning the full impacts will only become apparent in the longer term (OECD, 2018[15]).There is currently dedicated research and development into new sustainable petrochemical production routes (from production to use and disposal of products), which may contribute to efforts to curb and stop the leakage of plastic pollution and other harmful chemical products into the ocean (IEA, 2018[16]). New initiatives have also emerged, such as the World Economic Forum's Global Plastic Action Partnership. However, much progress is still required to address the root of chemical pollution, in terms of both the research and development necessary to find potential alternatives that would be less damaging to the environment and changing current production and consumption practices (e.g. building on the circular economy concept) (OECD, 2018[14]). In this context, developing countries experience particular pressures and vulnerabilities, including the generation and management of waste and the presence of marine plastic debris, often originating from distant waters.

Overfishing, by-catch and IUU fishing and over-exploitation of other natural resources

According to the Food and Agriculture Organization, 34.2% of the fish stocks in the world's marine fisheries were classified as overfished in 2017, with the maximally sustainably fished stocks accounting for 59.6% of the total number of assessed stocks (FAO, 2020[17]). IUU fishing exacerbates overfishing and is associated with significant impacts, from both an economic and a food security perspective. Estimating the magnitude of IUU fishing and its many social impacts (e.g. slavery on ships) is complex and depends on many factors, such as the type of fishery, the geographic location and the availability of information. Policy instruments to deter and combat IUU are described in Chapter 4. Overexploitation of other natural resources such as shellfish and other organisms is also causing damage to the marine environment (IPBES, 2019[7]).

Habitat degradation

Habitat destruction along coastlines and in the ocean results from harmful fishing practices such as dynamite fishing or improper trawling; poor land use practices in agriculture, coastal development and forestry sectors; other human activities such as mining, dredging and anchoring; and tourism and coastal encroachment. For example, logging and vegetation removal can introduce sediments from soil erosion. Harbour development and other land-based activities (such as shrimp aquaculture) can lead to the destruction of mangroves, which serve as nurseries for species of fish and shellfish and provide flood protection. Poor shipping practices and coastal tourist activities such as snorkelling, boating and scuba diving come in direct contact with fragile wetlands and coral reefs, consequently damaging marine habitats and degrading the ecosystem services they provide (OECD, 2017[5]).

Invasive alien species

Another serious threat to the marine environment is the introduction of non-native marine species to marine ecosystems to which they do not belong. Most of these alien species had been rapidly introduced to a different habitat through ballast water from commercial shipping operations across the oceans. These foreign organisms are responsible for severe environmental impacts, such as altering native ecosystem by disrupting native habitats, extinction of some marine flora and fauna, decreased water quality, increasing competition and predation among species, and spread of disease (OECD, 2017[5]). Considering these situations, the International Maritime Organization has made international efforts to address the transfer of invasive aquatic species through shipping, as illustrated by the adoption of the International Convention for the Control and Management of Ships' Ballast Water in February 2004. The Convention entered into force in September 2017.

The degradation of marine ecosystems is extending beyond ecologically and economically sustainable thresholds. One of the underlying reasons is that many of the services provided by marine and coastal ecosystems – such as coastal protection, fish nursery, water purification, marine biodiversity and carbon sequestration (Table 1.1) – are not reflected in the prices of traditional goods and services on the market (and hence referred to as non-market values). While there is often a lack of scientific information to clearly understand the complex links between these marine ecosystem services and their economic value, this undervaluation of marine ecosystem services results in under-investment in their conservation, sustainable use and restoration and lost opportunities for economic growth and poverty reduction, both now and for the future. Box 1.1 presents examples of the costs of inaction.

> Box 1.1. The cost of inaction on halting marine ecosystem degradation
>
> - The estimated economic losses due to overfishing amount to USD 83 billion annually, according to the World Bank (2017[18]).
> - The loss of tropical reef cover due to ocean acidification will cause damages of between USD 528 billion and USD 870 billion (year 2000 value) by 2100, according to Brander and Eppink (2012[19]).
> - The total estimated cost of coastal protection, relocation of people and loss of land to sea-level rise ranges from about USD 200 billion (for a 0.5-metre increase of sea level) to five times that amount, or USD 1 trillion, (for a rise of 1 metre) and to about USD 2 trillion (for an increase of 2 metres) (Nicholls and Cazenave, 2010[20]).
> - In the absence of proactive mitigation measures, climate change will increase the cost of damage to the ocean by an additional USD 322 billion per year by 2050 (Noone, Sumaila and Diaz, 2012[21]).
>
> Source: Adapted from OECD (2017[5]), *Marine Protected Areas: Economics, Management and Effective Policy Mixes*, https://doi.org/10.1787/9789264276208-en.

The sustainable ocean economy is multidimensional and rests on three pillars: economic, environmental and social

If managed sustainably, the ocean has the capacity to regenerate, be more productive, resilient, and support more equitable societies. A sustainable ocean economy needs to integrate all dimensions of sustainability – economic, social and environmental – in keeping with the 2030 Agenda and the SDGs.

Sustainable ocean economies have consequences that extend beyond environment considerations and a single ocean-focused SDG. They are intrinsically connected with many other SDGs (Le Blanca, Freire and Vierros, 2017[22]) and key to achieving economic, social and environmental sustainability. For instance, sustainable aquaculture and mariculture can contribute to feeding a growing and wealthier global population, advancing progress on SDG 1 (no poverty) and SDG 2 (zero hunger) by 2030. New opportunities from marine-based renewable energy can create new jobs and steer the global energy mix towards a greater share of clean energy, contributing to achievement of SDG 8 (decent work and economic growth) and SDG 13 (climate action). SDG 13 achievement can also be advanced through actions to preserve the ocean's capacity to regulate the climate and support biodiversity (Hoegh-Guldberg et al., 2019[23]), with a positive impact on climate migration and peace and stability (SDG 16). Table 1.2 illustrates these interlinkages, that conservation and sustainable use of the ocean can open new pathways towards the broader goal of inclusive, sustainable development.

Table 1.2. SDG 14 can contribute to and is strongly linked with other SDGs

SDG	Impact of SDG 14	Impact on SDG 14	Description of linkage
SDG1 – No poverty	X		Healthy oceans are enablers of poverty alleviation, especially in many SIDS and LDCs where fishery, aquaculture and tourism are key sources of livelihoods for coastal populations
SDG 2 – Food security	X		Sustainable aquaculture and innovative solutions on mariculture can significantly help meet the increasing global demand for food if managed sustainably
	X		Reducing ocean pollution and overfishing can put an end to fish stocks depletion and negative impacts on the quality of edible fish and other marine products
SDG 3 – Health and well-being	X		Addressing pollution of coastal areas and marine resources can curb its negative impacts on health and well-being
	X		The ocean is part of human well-being for cultural and recreational value people attach to it. Marine biodiversity provides a multitude of animals, algae and bacteria that can support the development of new medicines and vaccines
SDG 4 – Quality education		X	Ocean literacy is key to ensure awareness and scale up action to protect and conserve marine ecosystems
SDG 5 – Gender equality	X		Sustainable development of ocean-based industries can create new economic opportunities for women
SDG 6 – Water		X	Addressing wastewater (industrial and residential) and agricultural runoff reduces ocean pollution
		X	Wetlands protect water quality by trapping sediments and retaining excess nutrients and other pollutants such as heavy metals that may otherwise end up in the sea
SDG 7 – Affordable and clean energy		X	Reducing GHG emissions can contribute to ocean health by reducing ocean warming and acidification.
		X	Energy infrastructures in coastal and marine environment can have negative impacts on ocean health and represent a threat for marine ecosystems
SDG 8 – Economic growth and employment	X	X	Expansion in traditional and emerging ocean-based economic activities can help boost employment (e.g. in offshore wind energy, marine aquaculture, fish processing and port activities)
SDG 9 – Industrialisation and infrastructure	X	X	Industrial by-products and waste (e.g. heavy metals, chemicals, particulate matters) pollute oceans. On the other hand, efforts to improve the quality of infrastructure and planning for industrialisation could have large positive impacts on coastal areas currently detrimentally impacted by industry
SDG10 – Reduced inequality	X		Enhancing the quality of jobs and the social sustainability of ocean-related sectors can boost socio-economic inclusion
SDG 13 – Climate change		X	Pollution acts with other stressors to hamper the resilience of ecosystems to climate change
	X	X	The ocean's capacity to regulate the climate can be enhanced by conserving and enhancing the ocean carbon sink
SDG 15 – Terrestrial ecosystems		X	Better management of terrestrial ecosystems can reduce ocean pollution
SDG 16 – Peaceful and inclusive societies	X		Conserving the ocean's capacity to regulate the climate can prevent large climate migration flows
SDG 17 – Partnership to achieve the SDGs		X	The transboundary nature of the ocean and the global connectivity of marine ecosystems requires the establishment of effective partnerships between different countries and actors

Note: For a detailed analysis of interlinkages across the SDGs, see International Council on Science (2017[24]), A Guide to SDG Interactions: From Science to Implementation, https://council.science/wp-content/uploads/2017/05/SDGs-Guide-to-Interactions.pdf.

As the linkages show, the conservation and sustainable use of the ocean is a complex, cross-cutting issue. It calls for coherent governance approaches and proactive public policies to shape and create markets, fostering new products and business models as is discussed in Chapters 3 and 4. The core concept of sustainability of the ocean economy, much as it is of sustainable development as a whole, is multidimensional and rests on three, interrelated, pillars: economic, environmental and social.

Economic dimension

The ocean brings tangible economic benefits to actors around the world (OECD, 2016[4]). Economic activities from ocean-based industries are those that take place on or in the ocean; use inputs derived from the marine environment; produce goods and services for use on or in the ocean; and/or would not take place were they not located in proximity to the ocean (OECD, 2020[25]).

There is a wide range of formal, ocean-based industries (e.g. marine fishing, marine aquaculture, marine fish processing, shipbuilding, maritime passenger transport and maritime freight transport), as well as many informal but crucial subsistence and artisanal activities (e.g. artisanal fisheries). Many of these activities provide revenue and employment for millions of people around of the world. More than 3 billion people already rely on the ocean for their livelihoods, and an expected global population of 9 billion by 2050 will increase pressure to produce more food, energy and jobs from the ocean. Tracking the contribution of these ocean activities to the overall economy is needed to raise public awareness of the importance of the ocean, offering higher visibility to both investment opportunities in these sectors and to the crucial sustainability problems that demand action at many levels (OECD, 2019[26]). In order to improve the measurement of these ocean-based economic activities at national and international levels, and enhance comparability across sectors and across countries, greater standardisation of measurement approaches will be needed, with increased links to national accounting frameworks (OECD, 2020[25]).

In addition to these ocean-based economic activities, the marine environment itself brings many benefits. The two are interdependent in that much activity associated with ocean-based industry is derived from marine ecosystems (via the goods and services they provide), while most ocean-based industrial activity impacts marine ecosystems (OECD, 2016[4]).

Environmental dimension

Marine ecosystems provide invaluable life support functions. Most of these values are not reflected in market prices however, which leads to significant negative externalities. Integrating the environmental pillar into sustainability requires that these inherent values are reflected in decision-making processes. This involves understanding the total economic value provided by ecosystems (i.e. direct and indirect use values and non-use values) and taking these into account in cost-benefit analysis.

According to the OECD definition that was developed for statistical measurement purposes, the ocean economy is the sum of the economic activities of ocean-based industries plus the assets, goods and services provided by marine ecosystems (OECD, 2016[4]). This definition encompasses all natural assets and ecosystem services that the ocean provides such as habitat provisioning and CO_2 absorption. An example is coral reefs. They provide shelter and habitat for fish nurseries and unique genetic resources, while at the same time providing recreational value for maritime tourism.

At the international level, progress is being made to develop natural capital accounting frameworks, including for marine ecosystem services, within the System of National Accounts (SNA). The System of Environmental Economic Accounting (SEEA) was adopted as an international standard in 2012. The SEEA extends the SNA accounting framework to include environmental issues that are biophysical in nature. The SEEA Experimental Ecosystem Accounting framework, developed more recently, goes further and describes how to account for ecosystem assets and their provisioning, regulatory and cultural services in both physical and monetary terms. While earlier work focused more heavily on terrestrial ecosystems, progress now is being made on including marine ecosystems (OECD, 2019[26]). *The Changing Wealth of Nations 2020* covers new assets including the ocean (fisheries, mangroves and coral reef) for example (World Bank, 2020[27]).

Social dimension

Sustainable development intersects and conveys the often conflicting objectives of economic and social development and of environmental protection (OECD, 2019[28]). The social dimension of sustainable development encompasses many factors, among them poverty alleviation, food security, inclusive growth and intergenerational equity.

The ocean economy is an integral part of the social fabric of much of the world's people, particularly in developing countries. Subsistence fishery not only provides some coastal populations with food. It also contributes to cultural and societal cohesion. The ocean economy also is a key contributor to alleviating poverty and food security: seafood is the primary source of animal protein in the diets of approximately 1 billion people, mostly in developing countries (OECD/FAO, 2019[29]).

Major challenges remain, particularly for vulnerable communities, in terms of inclusive growth (i.e. economic growth that is distributed fairly across society and creates opportunities for all), unequal access to marine resources, and the impacts of pollution on some populations. The intergenerational effects are also important, as monitoring the stocks and trends of existing marine resources is not only crucial for today's generations but constitutes a step towards understanding their prospects for the future well-being of the next generations. Beyond GDP growth, the well-being that may be derived from the ocean economy is also not yet well defined or measured.

A sustainable ocean economy will only be achieved, in principle, when the three constitutive economic, environmental and social dimensions are well understood, with robust evidence such as:

- Statistical measurement of industries and ecosystems services,
- Adequate environmental impact assessments of economic activities,
- And social well-being indicators.

Institutional and governance mechanisms can contribute to empowering people and moving towards more inclusiveness and equality (OECD, 2019[30]). The three dimensions of a sustainable ocean economy can then be mutually reinforcing, when a policy coherence framework, and its control and enforcement measures, such as regulations and economic instruments (e.g. taxes), are actually in place.

References

Bijma, J. et al. (2013), "Climate change and the oceans – What does the future hold?", *Marine Pollution Bulletin*, Vol. 74/2, pp. 495-505, http://dx.doi.org/10.1016/j.marpolbul.2013.07.022. [9]

Brander, L. and F. Eppink (2012), *The Economics of Ecosystems and Biodiversity for Southeast Asia (ASEAN TEEB)*, http://lukebrander.com/wp-content/uploads/2013/07/ASEAN-TEEB-Scoping-Study-Report.pdf. [19]

Díaz, S. et al. (eds.) (2019), *Global assessment report on biodiversity and ecosystem services of the Intergovernmental Science-Policy Platform on Biodiversity and Ecosystem Services (IPBES)*, Intergovernmental Science-Policy Platform on Biodiversity and Ecosystem Services (IPBES) Secretariat, Bonn, https://doi.org/10.5281/zenodo.3553579. [7]

Diez, S. et al. (2019), *Marine Pollution in the Caribbean: Not a Minute to Waste*, World Bank Group, Washington, DC, http://documents1.worldbank.org/curated/en/482391554225185720/pdf/Marine-Pollution-in-the-Caribbean-Not-a-Minute-to-Waste.pdf. [13]

FAO (2020), *In Brief: The State of World Fisheries and Aquaculture 2020 - Sustainability in Action*, Food and Agriculture Organization (FAO), Rome, http://dx.doi.org/10.4060/ca9231en. [17]

GESAMP (2001), *Protecting the Oceans from Land-based Activities*, Joint Group of Experts on the Scientific Aspects of Marine Environmental Protection (GESAMP), London, http://www.gesamp.org/publications/protecting-the-oceans-from-land-based-activities. [12]

GOC (2014), *From Decline to Recovery: A Rescue Package for the Global Oceans*, Global Oceans Commission, https://www.mckinsey.com/~/media/mckinsey/dotcom/client_service/Sustainability/PDFs/From_decline_to_recovery_A_rescue_package_for_the_global_ocean.ashx. [11]

Griggs, D., M. Nilsson and A. Stevance (eds.) (2017), *A Guide to SDG Interactions: From Science to Implementation*, http://dx.doi.org/10.24948/2017.01. [24]

IEA (2018), *The Future of Petrochemicals: Towards More Sustainable Plastics and Fertilisers*, International Energy Agency, Paris, https://dx.doi.org/10.1787/9789264307414-en. [16]

International Chamber of Shipping (2020), *Key facts: Shipping and world trade (webpage)*, https://www.ics-shipping.org/shipping-facts/key-facts (accessed on 16 June 2020). [3]

IPCC (2019), *The Ocean and Cryosphere in a Changing Climate*, ntergovernmental Panel on Climate Change (IPCC), Geneva, https://www.ipcc.ch/report/srocc/ (accessed on 16 June 2018). [6]

Le Blanca, D., C. Freire and M. Vierros (2017), *Mapping the linkages between oceans and other Sustainable Development Goals: A preliminary exploration*, United Nations Department of Social and Economic Affairs, New York, https://sustainabledevelopment.un.org/content/documents/12468DESA_WP149_E.pdf. [22]

Nicholls, R. and A. Cazenave (2010), "Sea-level rise and its impact on coastal zones", *Science*, Vol. 328/5985, pp. 1517-1520, http://dx.doi.org/10.1126/science.1185782. [20]

Noone, K., R. Sumaila and R. Diaz (2012), *Valuing the Ocean: Preview Summary*, Stockholm Environment Institute, Stockholm, https://www.sei.org/publications/valuing-the-ocean-preview-summary/. [21]

OECD (2020), *A Blueprint for Improved Measurement of the International Ocean Economy*, OECD Publishing, Paris. [25]

OECD (2019), *Development Co-operation Report - Profiles (webpage)*, http://www.oecd.org/dac/development-cooperation-report/#profiles (accessed on 12 November 2019). [32]

OECD (2019), *OECD Recommendation on Policy Coherence for Sustainable Development*, OECD Publishing, Paris, https://legalinstruments.oecd.org/en/instruments/OECD-LEGAL-0381. [28]

OECD (2019), *Policy Coherence for Sustainable Development 2019: Empowering People and Ensuring Inclusiveness and Equality*, OECD Publishing, Paris, https://dx.doi.org/10.1787/a90f851f-en. [30]

OECD (2019), *Rethinking Innovation for a Sustainable Ocean Economy*, OECD Publishing, Paris, https://dx.doi.org/10.1787/9789264311053-en. [26]

OECD (2018), *Improving Markets for Recycled Plastics: Trends, Prospects and Policy Responses*, OECD Publishing, Paris, https://dx.doi.org/10.1787/9789264301016-en. [14]

OECD (2018), "Improving plastics management: Trends, policy responses, and the role of international co-operation and trade", *OECD Environment Policy Papers*, No. 12, OECD Publishing, Paris, https://dx.doi.org/10.1787/c5f7c448-en. [15]

OECD (2017), *Marine Protected Areas: Economics, Management and Effective Policy Mixes*, OECD Publishing, Paris, https://doi.org/10.1787/9789264276208-en (accessed on 1 May 2018). [5]

OECD (2016), *The Ocean Economy in 2030*, OECD Publishing, Paris, https://dx.doi.org/10.1787/9789264251724-en. [4]

OECD DAC (2020), *DAC List of ODA Recipients*, http://www.oecd.org/dac/financing-sustainable-development/development-finance-standards/DAC-List-of-ODA-Recipients-for-reporting-2020-flows.pdf. [33]

OECD/FAO (2019), *OECD-FAO Agricultural Outlook 2019-2028*, OECD Publishing, Paris/Food and Agriculture Organization of the United Nations, Rome, https://dx.doi.org/10.1787/agr_outlook-2019-en. [29]

UN (2017), *Our ocean, our future: Call for Action*, United Nations General Assembly, New York, https://www.un.org/ga/search/view_doc.asp?symbol=A/RES/71/312&Lang=E. [2]

UN (2015), *Transforming Our World: The 2030 Agenda for Sustainable Development*, United Nations General Assembly, New York, https://sustainabledevelopment.un.org/post2015/transformingourworld/publication. [1]

UNDP (2012), *Catalysing Ocean Finance. Volume 1: Transforming Markets to Restore and Protect the Global Ocean*, United Nations Development Programme, https://www.undp.org/content/undp/en/home/librarypage/environment-energy/water_governance/ocean_and_coastalareagovernance/catalysing-ocean-finance.html. [10]

UNEP (2011), *Towards a Green Economy: Pathways to Sustainable Development and Poverty Eradication*, United Nations Environment Programme, https://sustainabledevelopment.un.org/content/documents/126GER_synthesis_en.pdf. [8]

World Bank (2020), *The Changing Wealth of Nations*, https://www.worldbank.org/en/topic/environment/publication/changing-wealth-of-nations. [27]

World Bank (2020), *World Bank Country and Lending Groups (webpage)*, https://datahelpdesk.worldbank.org/knowledgebase/articles/906519-world-bank-country-and-lending-groups. [31]

World Bank (2017), *The Sunken Billions Revisited: Progress and Challenges in Global Marine Fisheries*, World Bank, Washington, DC, http://dx.doi.org/10.1596/978-1-4648-0919-4. [18]

World Resources Institute, W. (ed.) (2019), *The Ocean as a Solution to Climate Change: Five Opportunities for Action*, http://www.oceanpanel.org/climate. [23]

Annex 1.A. Country groupings

Although there is no established convention designating countries as either developed or developing, but the most widely used classification is based on income groups of low to high-income countries and on other structural elements. This report focuses on countries below the high- income threshold, which broadly corresponds to those that are eligible to receive development assistance (ODA). Its findings pertain to specific country groupings within this set of diverse countries such as least developed countries and small island developing states. The definitions underlying these classifications are briefly explained below.

Classification by income groups: The World Bank classifies countries into four income groups based on their gross national income (GNI) per capita in current USD. Currently, the breakdown is high (80 countries), upper middle (60 countries), lower middle (47 countries) and low income (31 countries), (World Bank, 2020[31]).

Least developed countries (LDC) category: The United Nations (UN) General Assembly officially established the LDC category in 1971 to identify countries with serious structural impediments to development beyond those suggested by GNI alone. LDCs currently comprise 47 low-income and lower middle-income countries that are highly vulnerable to economic and environmental shocks. They are home to about 880 million people, 12% of the world population, but account for less than 2% of world gross domestic product (GDP) and about 1% of world trade. The Committee for Development Policy, a subsidiary body of the UN Economic and Social Council, is mandated to review the countries categorised as LDCs every three years and monitor their progress after they graduate out of the category.

Eligibility to official development assistance (ODA): Least developed countries and all other low and middle-income countries, based on GNI per capita (atlas method) as published by the World Bank, are eligible to access international support such as grants and concessional loans, with the exception of Group of Eight members and members of the European Union. The OECD Development Assistance Committee (DAC) maintains the List of ODA Recipients, which is updated every three years. To graduate from the DAC list, a country must exceed the high-income threshold set by the World Bank for three consecutive years. A country can therefore be classified as high income, yet still be eligible to receive ODA (OECD, 2019[32]).

Small island developing states: A number of lists exist of small island developing states (SIDS), among them lists established by the UN Office of the High Representative for the Least Developed Countries, Landlocked Developing Countries and Small Island Developing States (UN-OHRLLS) comprising 52 SIDS, of which 38 are UN member states); the Alliance of Small Island States, comprising 39 SIDS; and the UN Conference on Trade and Development, comprising 29 SIDS. The World Bank Group defines small states as countries that have a population of 1.5 million or less or are members of the World Bank Group Small States Forum. Currently, 50 small landlocked and coastal states fit this definition, including 27 of the 35 ODA-eligible SIDS considered in this report. Given that a main theme of this report is the role of development co-operation, it focuses on the 34 small island developing states that are currently ODA-eligible, which comprise: 9 LDCs, 5 lower middle-income countries and 20 upper middle-income countries.

List of countries and territories by income groupings: The World Bank uses four income groupings: low, lower middle, upper middle and high income. Income is measured as GNI per capita, in USD converted from local currency. Following is the current World Bank (2020[31]) list by income:

- Low-income economies (USD 1 025 or less) : Afghanistan, Benin, Burkina Faso, Burundi, Central African Republic, Chad, Democratic Republic of the Congo, Eritrea, Ethiopia, Gambia, Guinea, Guinea-Bissau, Sierra Leone, Haiti, Democratic People's Republic of Korea, Liberia, Madagascar,

Malawi, Mali, Mozambique, Nepal, Niger, Rwanda, Somalia, South Sudan, Syrian Arab Republic, Tajikistan, United Republic of Tanzania, Togo, Uganda and Yemen.

- Lower-middle income economies (USD 1 026 to USD 3 995): Angola, Bangladesh, Bhutan, Plurinational State of Bolivia, Cabo Verde, Cambodia, Cameroon, Comoros, Republic of the Congo, Côte d'Ivoire, Djibouti, Egypt, El Salvador, Kingdom of Eswatini, Ghana, Honduras, India, Indonesia, Kenya, Kiribati, Kyrgyzstan, Lao People's Democratic Republic, Lesotho, Mauritania, Federated States of Micronesia, Republic of Moldova, Mongolia, Morocco, Myanmar, Nicaragua, Nigeria, Pakistan, Papua New Guinea, Philippines, Sao Tome and Principe, Senegal, Solomon Islands, Sudan, Timor-Leste, Tunisia, Ukraine, Uzbekistan, Vanuatu, Viet Nam, West Bank and Gaza Strip, Zambia, Zimbabwe.

- Upper-middle-income economies (USD 3 996 to USD 12 375): Albania, Algeria, American Samoa, Argentina, Armenia, Azerbaijan, Belarus, Belize, Bosnia and Herzegovina, Botswana, Brazil, Bulgaria, People's Republic of China, Colombia, Costa Rica, Cuba, Dominica, Dominican Republic, Equatorial Guinea, Ecuador, Fiji, Gabon, Georgia, Grenada, Guatemala, Guyana, Islamic Republic of Iran, Iraq, Jamaica, Jordan, Kazakhstan, Kosovo, Lebanon, Libya, Malaysia, Maldives, Marshall Islands, Mauritius, Mexico, Montenegro, Namibia, Nauru, Republic of North Macedonia, Paraguay, Peru, Romania, Russian Federation, Samoa, Serbia, Sri Lanka, South Africa, Saint Lucia, Saint Vincent and the Grenadines, Suriname, Thailand, Tonga, Turkey, Turkmenistan, Tuvalu and Bolivarian Republic of Venezuela.

- High-income economies (USD 12 376 and more) Andorra, Antigua and Barbuda, Aruba, Australia, Austria, Bahamas, Bahrain, Barbados, Belgium, Bermuda, British Virgin Islands, Brunei Darussalam, Canada, Cayman Islands, Bailiwick of Guernsey, Bailiwick of Jersey, Chile, Croatia, Curaçao, Cyprus2, Czech Republic, Denmark, Estonia, Faroe Islands, Finland, France, French Polynesia, Germany, Gibraltar, Greece, Greenland, Guam, Hong Kong, China, Hungary, Iceland, Ireland, Isle of Man, Israel, Italy, Japan, Korea, Kuwait, Latvia, Liechtenstein, Lithuania, Luxembourg, Macau, China, Malta, Monaco, Netherlands, New Caledonia, New Zealand, Northern Mariana Islands, Norway, Oman, Palau, Panama, Poland, Portugal, Puerto Rico, Qatar, San Marino, Saudi Arabia, Seychelles, Singapore, Sint Maarten, Slovak Republic, Slovenia, Spain, Saint Kitts and Nevis, Saint Martin, Sweden, Switzerland, Chinese Taipei, Trinidad and Tobago, Turks and Caicos Islands, United Arab Emirates, United Kingdom, United States, Uruguay and United States Virgin Islands.

- ODA-eligible SIDS by income group and by region are shown in Annex Figure 1.A.1.

Annex Figure 1.A.1. ODA-eligible SIDS by income group and by region

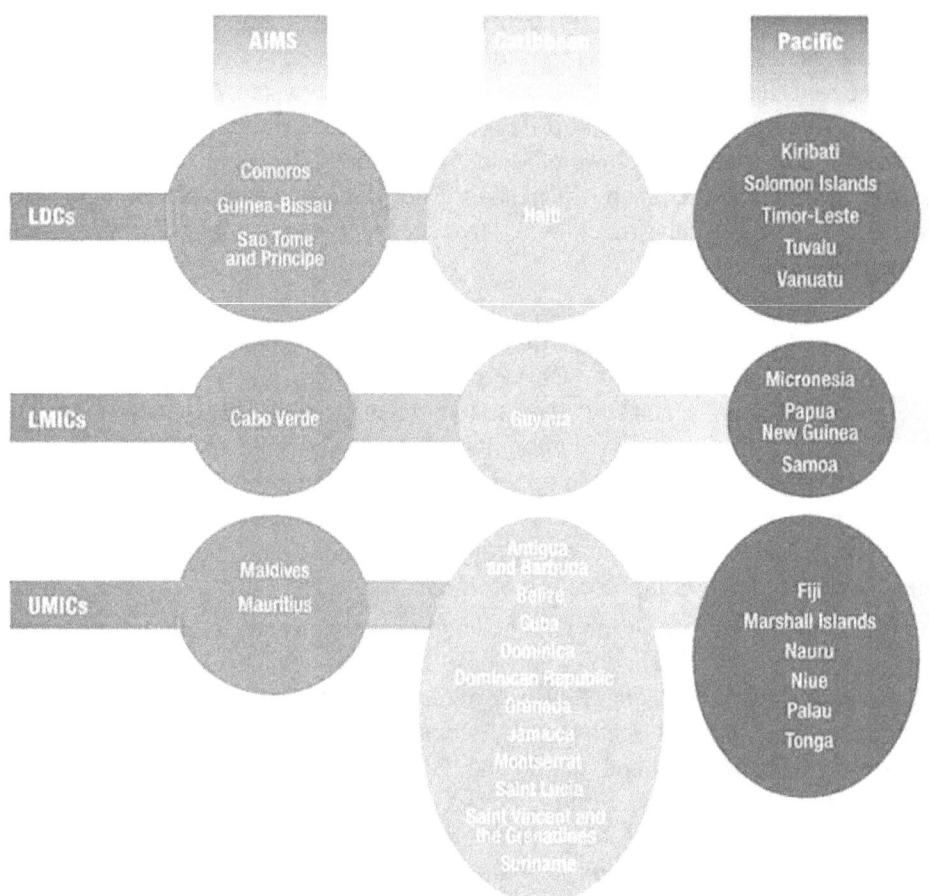

Source: Adapted from OECD DAC (2020[33]), List of ODA Recipients, http://www.oecd.org/dac/financing-sustainable-development/development-finance-standards/DAC-List-of-ODA-Recipients-for-reporting-2020-flows.pdf and World Bank (2020[31]), World Bank Country and Lending Groups (webpage), https://datahelpdesk.wodrldbank.org/knowledgebase/articles/906519-world-bank-country-and-lending-groups.

Notes

[1] For the definition of developing countries, see Annex 1.A.

[2] Note by Turkey: The information in this document with reference to "Cyprus" relates to the southern part of the Island. There is no single authority representing both Turkish and Greek Cypriot people on the Island. Turkey recognises the Turkish Republic of Northern Cyprus (TRNC). Until a lasting and equitable solution is found within the context of United Nations, Turkey shall preserve its position concerning the "Cyprus" issue.

2 Developing countries and the ocean economy: Key trends

This chapter provides an overview of selected trends in the ocean economy in developing countries, based in part on new OECD experimental ocean-based industries datasets. The following sectors of particular interest to developing countries are introduced and include marine fishing and aquaculture, coastal and marine tourism, extractive industries (e.g. oil and gas, sea-bed mining), transport and logistics industries (freight and passenger transport), shipbuilding industries, renewable energy and marine biotechnologies. All these economic activities bring long-term sustainability challenges. In view of the COVID-19 crisis' enduring economic impacts on the ocean economy, the evidence provided here should contribute to providing a useful baseline for purposes of tracking the positioning of some countries and planning for recovery and new developments. As OECD measurement activities continue in partnership with the international community and ocean industry players, additional and improved economic evidence will be developed.

Positioning of developing countries in ocean-based industries: Trends across country groupings

Not all developing countries have the same connections to the ocean. Geography, history, culture and economic development are all crucial elements that determine the links between countries and the ocean economy. While coastal states and small island developing states (SIDS) have embraced artisanal fisheries for centuries, the development of commercial ports activities in some of these same countries may be relatively recent and driven by changes in regional and international trade. Still, the ocean economy is already important for many countries around the world in terms of jobs and revenues, and particularly so for the many developing countries where a larger proportion of economic activity stems from ocean resources and their use. Measuring the value of ocean activities is therefore important to inform policy- and decision-makers in the public and private sectors, so as to foster improved ocean governance, and enhance sustainable practices.

The economic value of ocean-based industries

The OECD (2016[1]) report, *The Ocean Economy in 2030*, focussed on the task of measuring the economic contribution of ocean-based industries worldwide, providing for the first time a broad definition of the ocean economy based on the interactions among economic activities and the marine environment (Chapter 1). As a starting point, it estimated that in 2010, ten ocean-based industries produced value added equivalent to 2.5% of global gross domestic product (GDP), a total of USD 1.5 trillion, and full-time employment for approximately 30 million people. Offshore oil and gas accounted for about one third of total value added of the ocean-based industries, followed by maritime and coastal tourism (26%), ports (13%), and maritime equipment (11%). The other industries accounted for shares of 5% or less (industrial fish processing, shipping, shipbuilding and repair, industrial capture fisheries, aquaculture, and offshore wind). Industrial capture fisheries were one of the smallest in value added, although the available data did not take into account the importance of artisanal fisheries. In terms of employment, the ocean-based industries contributed some 31 million direct, full-time jobs in 2010, (roughly equal to France's entire labour force that year). The largest employers (not including all artisanal fisheries) were industrial capture fisheries and maritime and coastal tourism.

Projecting to 2030, the report estimated that these ten ocean-based industries would reach, conservatively, a gross value added of about USD 3 trillion (roughly equivalent to the size of the German economy in 2010). Before the impacts of the COVID-19 pandemic began to spread, tourism development, including the cruise industry, was expected to make up the largest share (26%), followed by offshore oil and gas exploration and production (21%), port activities (16%), marine equipment (10%), fish processing (10%), and offshore wind (8%). The other industries were below 5%. Notable were the high rates of growth expected in some of the sectors (e.g. marine aquaculture, offshore wind and port activities). A majority of the workforce was expected to be working in the industrial capture fisheries sector and the maritime and coastal tourism industry. As noted, these are very conservative estimates, not least because several important activities in the ocean economy (e.g. marine business and finance, ocean surveillance, and marine biotechnology) were not included due to lack of data.

Since this initial assessment in 2016, developments in tourism and offshore oil and gas exploration, in particular, have accelerated, as discussed in more detail in the sector-specific sections of this report. The OECD, in co-operation with many countries, has also been undertaking further statistical analysis of ocean-based industries (Table 2.1) including original work on satellite accounts for the ocean economy (OECD, 2019[2]). Six ocean-based industries were the first focus of this analysis, starting in 2019, to produce long-term, experimental time series of value added and employment as a subset of the ocean economy. The six industries – marine fishing, marine aquaculture, marine fish processing, shipbuilding, maritime passenger transport and maritime freight transport – were selected based on the availability of datasets that could be populated for the period 2000-15 using official and internationally comparable sources. Many of these are of particular relevance for developing countries, and they interestingly differ in terms of their industrial structure, infrastructure needed and levels of qualification in human resources. The six ocean-based industries represented a total of approximately USD 376 billion in value added in 2015. Other ocean-based industries will be examined over the course of 2020 and into the future. A new OECD (forthcoming[3]) report on ocean economy measurement presents additional results and conceptual and methodological details.

The COVID-19 pandemic is having major economic impacts on countries around the world. Many ocean activities, especially those at the heart of global trade such as shipping as well as leisure activities such as coastal tourism, have been affected by measures undertaken to control the spread of the disease. The associated economic effects are considerable. Exactly how this disruption will impact the future of the ocean economy and the marine environment is not, as yet, clear. However, economic activity has slowed down markedly, and some time may pass before pre-crisis activity levels are reached again.

The looming, global economic downturn provoked by the pandemic has created a pervasive sense of uncertainty that may well persist in the short to medium term. Nonetheless, it will be important for countries to not lose sight of the longer-term opportunities that a sustainable ocean economy can offer. Recovery plans should especially anticipate and identify changes to demand and supply over the long run and consider how these changes might affect existing pressures on ocean ecosystems.

Table 2.1. Ocean-related economic activities

1	Marine fishing
2	Marine aquaculture
3	Maritime passenger transport
4	Maritime freight transport
5	Offshore extraction of crude petroleum and natural gas
6	Marine mining and dredging
7	Offshore industry support activities
8	Processing and preserving of marine fish, crustaceans and molluscs
9	Maritime ship, boat and floating structure building
10	Maritime manufacturing, repair and installation
11	Offshore wind and marine renewable energy
12	Maritime ports and support activities for maritime transport
13	Ocean scientific research and development
14	Marine and coastal tourism

Note: The list is evolving and may change in accordance with developments in ongoing OECD work on satellite accounts for the ocean economy.
Source: OECD (forthcoming[3]), A Blueprint for Improved Measurement of the International Ocean Economy, OECD Publishing, Paris.

The next subsections look at the value added generated in six selected ocean-based industries and compares these across different country income groupings (Annex 2.A) and across different regions. They also look at evidence from the new datasets to highlight trends pertaining to the ocean economy and developing countries. Two overarching trends transpire: First, lower income groups rely extensively on their natural assets to develop their ocean-based industries, and in some cases important shares of their GDP depend on these industries. And second, ocean-based industries' importance varies widely across different regions of the world, with some more advanced than others. Further insights in other sectors of the ocean economy are provided in the subsequent sections.

Lower income countries rely extensively on their natural assets to develop their ocean-based industries, and important shares of their GDP depend on these industries

A first focus on six ocean-based industries and comparable data on their value added provide some insights about the ocean economy and developing countries, as these industries constitute a subset of the ocean economy. Based on the new datasets, the economic positioning of developing countries varies widely across ocean-based industries. The six ocean-based industries include marine fishing, marine aquaculture, marine fish processing, shipbuilding, maritime passenger transport and maritime freight transport.

As shown in Figure 2.1, lower middle-income countries overall have a greater value added in marine fishing and fish processing, while high-income and upper middle-income countries generate the greatest value added in manufacturing and the transport sectors of shipbuilding, maritime freight transport and passenger transport.

Figure 2.1. Global value added in six ocean-based industries, by country income group, 2005-15

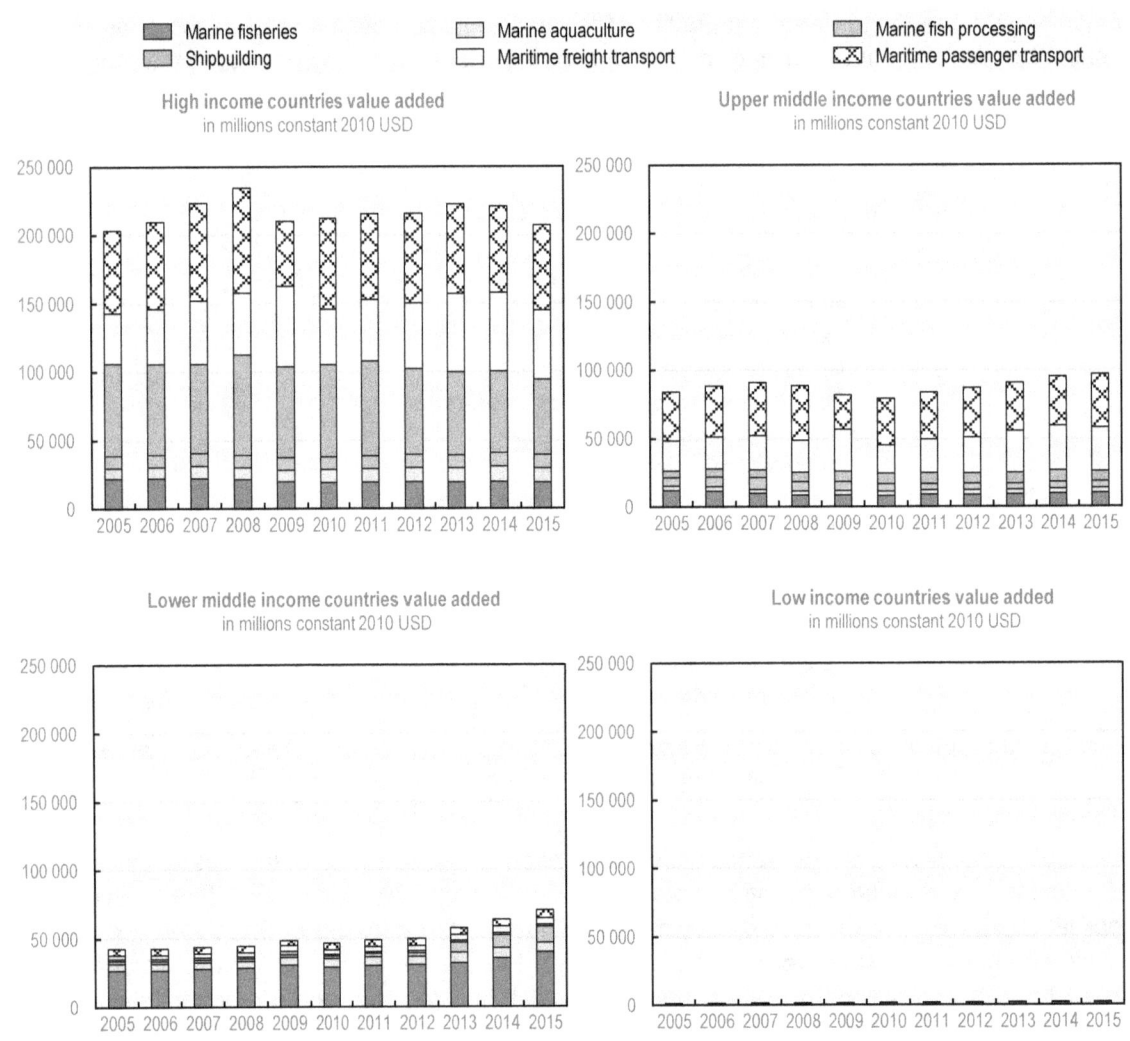

Note: The values presented in this figure are part of ongoing OECD experimental work in building up an ocean satellite account. Future estimates may vary for definitional reasons.
Source: OECD (2020[4]), Experimental Ocean-Based Industry Database, Directorate for Science, Technology and Innovation.

StatLink https://doi.org/10.1787/888934159145

The six ocean-based industries' share of GDP also varies across country income groups. In high-income and upper middle-income countries, they represent less than 2% in 2015. They represent much larger shares in 2015 in lower middle-income countries (more than 11%) and in low-income countries (just under 6%). While these averages mask hide large differences from country to country, this analysis highlights the relative importance of ocean-based industries, particularly marine fishing, in the economies of developing countries.

Figure 2.2. Share of GDP of value added from six ocean-based industries, by country income groups, 2005-15

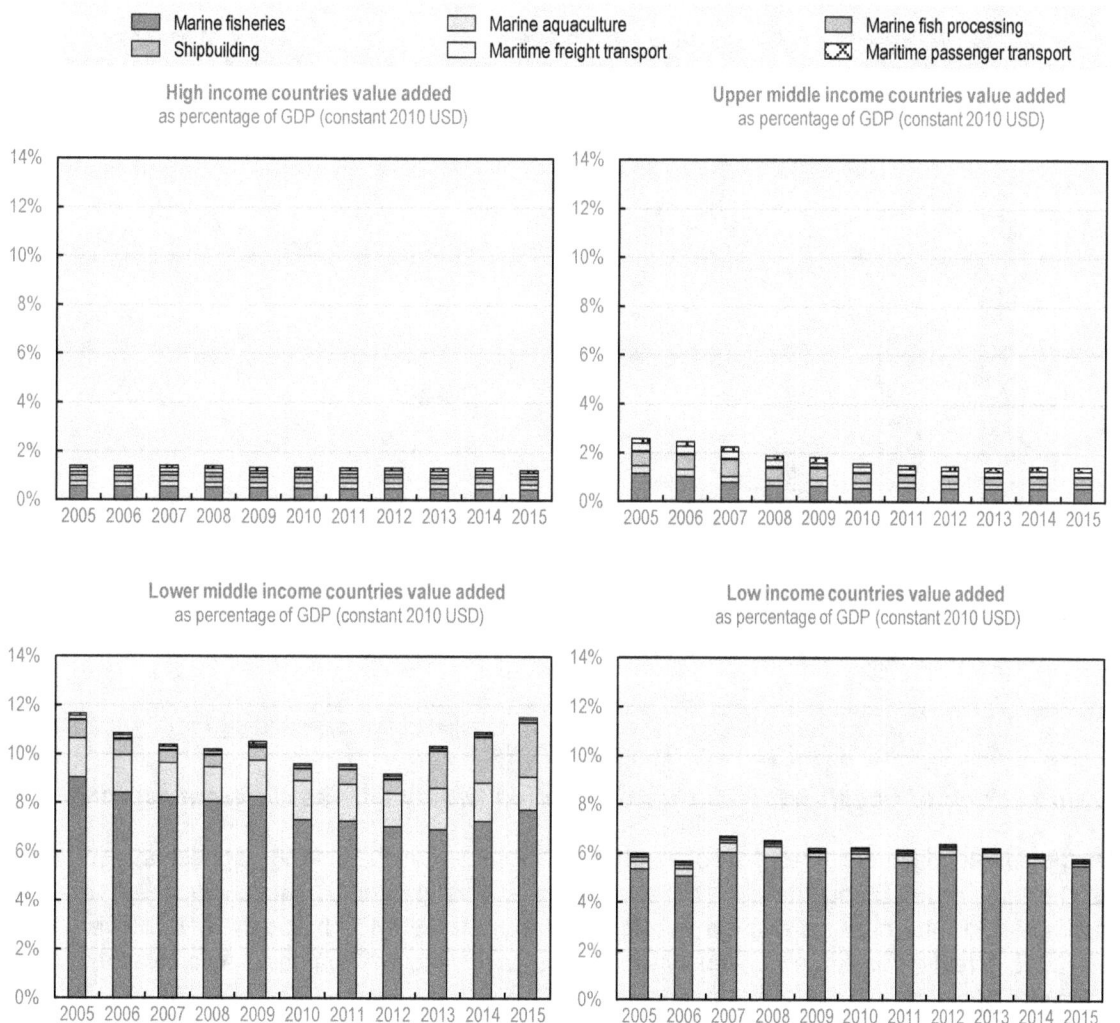

Note: The values presented here are part of ongoing OECD experimental work in building up an ocean satellite account. Future estimates may vary for definitional reasons.
Source: OECD (2020[4]), Experimental Ocean-Based Industry Database, Directorate for Science, Technology and Innovation.

StatLink https://doi.org/10.1787/888934159164

A final focus on the least developed countries (LDCs) category, which includes low-income and some lower middle- income countries, shows that marine fisheries accounts for the biggest share of total value added of the six ocean-based industries, representing 76% of the total (Figure 2.3). It accounts as well for a sizable proportion of employment in LDCs. These countries are some of the world's most vulnerable to economic and environmental shocks. Among the top ten LDCs in terms of total value added from the six ocean-based industries, Angola leads (Figure 2.3). This is due to the specificities of its marine fisheries and fish processing industries, which export high-value products (shrimp and fresh fish such as grouper and seabream) and import low-value fish for internal consumption (FAO, 2019[5]). The next highest are Bangladesh, Madagascar, Myanmar, Sierra Leone and Senegal.

Figure 2.3. Value added of selected ocean-based industries in least developed countries (LDCs)

Total value added in millions of constant 2010 USD (2015)

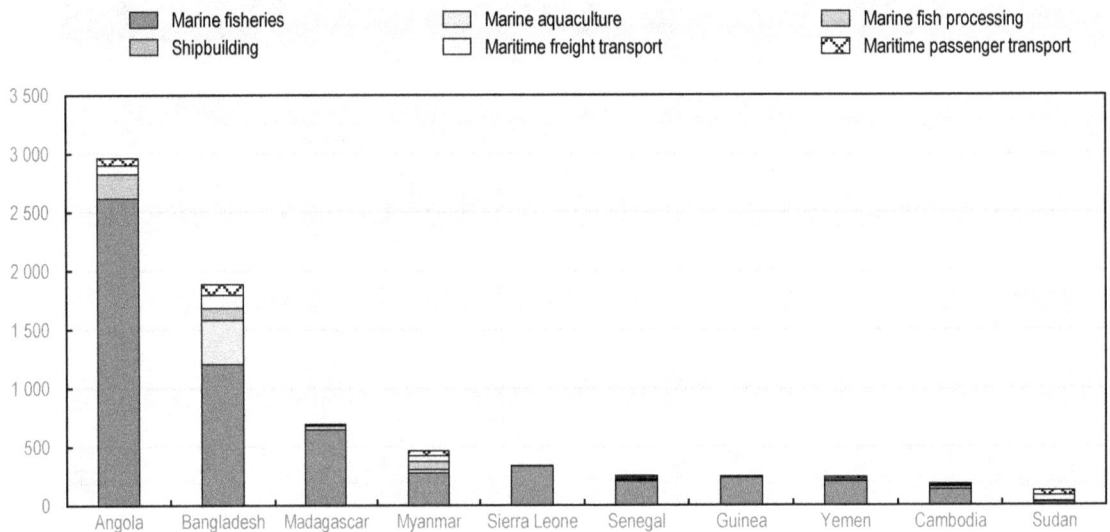

Note: The values presented here are part of ongoing OECD experimental work in building up an ocean satellite account. Future estimates may vary for definitional reasons.
Source: OECD (2020[4]), Experimental Ocean-Based Industry Database, Directorate for Science, Technology and Innovation.

StatLink https://doi.org/10.1787/888934159183

The value added of ocean-based industries varies widely across different regions

Total value added from the subset of six ocean-based industries over 2005-15 is highest, at approximately USD 200 billion, in OECD countries as a group. Among regional groupings, these industries' value added increased the most in East Asia and the Pacific, growing from USD 157 billion in 2010 to over USD 175 billion in 2015 in real terms, largely driven by China. The Association of Southeast Asian Nations (ASEAN) group (Brunei Darussalam, Cambodia, Indonesia, Lao People's Democratic Republic, Malaysia, Myanmar, Philippines, Singapore, Thailand and Viet Nam) is close to reaching USD 50 billion in value added from the ocean economy. The six ocean-based industries represent approximately USD 19 billion in value added in both Latin America and the Caribbean and in South Asia (Figure 2.4). A comparison of value added of these industries at five-year intervals between 2005 and 2015 shows growth in all regions of the world except Europe and Central Asia. It should also be noted that for many of these regions, marine and costal tourism and the offshore extraction of crude petroleum and natural gas are important ocean economy sectors but these are not included in the datasets.

Figure 2.4. Value added from six ocean-based industries, by region and country grouping

Total value added in millions of constant 2010 USD (2005, 2010 and 2015)

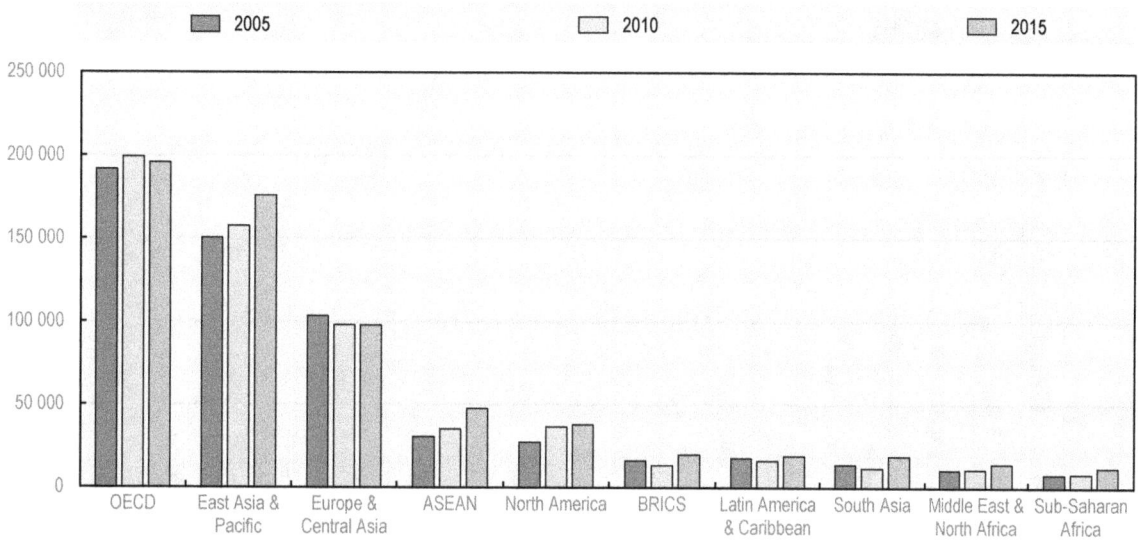

Note: Value added is from the ocean-based industries of marine fishing, marine aquaculture, marine fish processing, shipbuilding, maritime passenger transport and maritime freight transport. A country may belong to one or more country grouping. The values presented here are part of ongoing OECD experimental work in building up an ocean satellite account. Future estimates may vary for definitional reasons.
Source: OECD (2020[4]), Experimental Ocean-Based Industry Database, Directorate for Science, Technology and Innovation.

StatLink https://doi.org/10.1787/888934159202

Although the global value added for these six ocean based industries is highest overall in the OECD area, employment in these industries is among the lowest of the regional and country groupings (Figure 2.5). This is likely due to higher productivity in the higher-income countries. The East Asia and Pacific region has the largest number of jobs in these industries (over 20 million). The number of jobs in these ocean-based industries is slowly rising in the ASEAN grouping, from 8.4 million in 2005 to 9 million in 2015. In sub-Saharan Africa, marine activities accounted for about 1.9 million jobs in 2015. Inland fishing and inland aquaculture are also employment-intensive in parts of Africa but are not counted in these figures. These estimates are conservative as they do not take into account significant subsistence activities. A study by de Graaf and Garibaldi (2014[6]) estimated the gross added value of the fisheries and aquaculture sector in Africa, including many inland activities not covered in this report, at USD 24 billion in 2011 (i.e. 1.6% of the GDP of all African countries) and that about 12.3 million people were employed in the sector.

Figure 2.5. Employment in six ocean-based industries, 2005-15, by region and country grouping

Total employment in thousands of persons engaged (2005, 2010 and 2015)

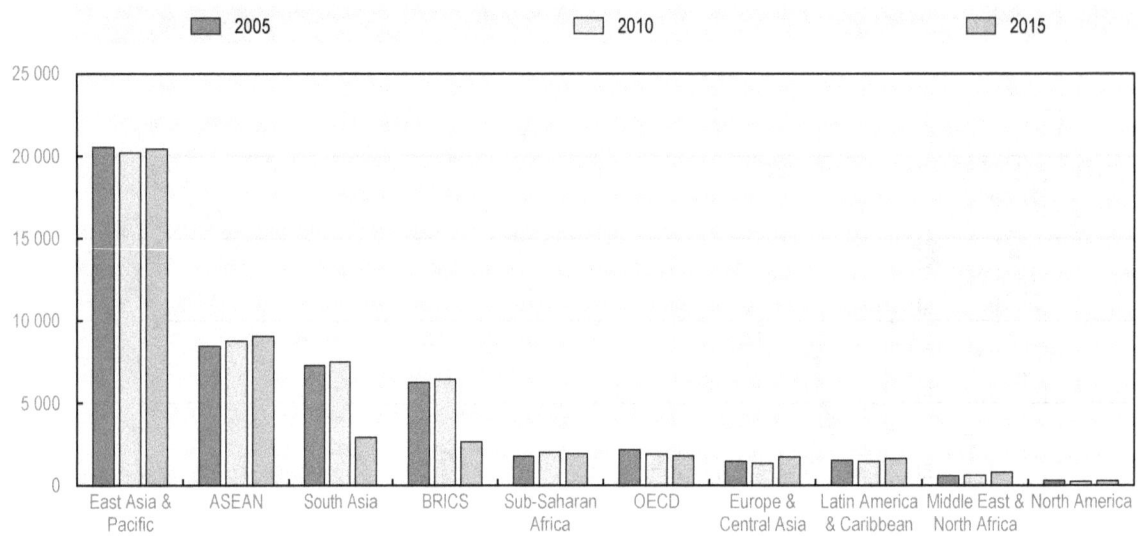

Note: The ocean-based industries measured are marine fishing, marine aquaculture, marine fish processing, shipbuilding, maritime passenger transport and maritime freight transport. A country may belong to one or more country grouping. The drop for the South Asia and BRICS groupings in 2015 are caused by statistical lags. The values presented here are part of ongoing OECD experimental work in building up an ocean satellite account. Future estimates may vary for definitional reasons.
Source: OECD (2020[4]), Experimental Ocean-Based Industry Database, Directorate for Science, Technology and Innovation.

StatLink https://doi.org/10.1787/888934159221

A more granular look at the role of the six ocean-based industries in the Caribbean region is illustrative. In terms of both value added and employment, marine fishing accounts for the largest share among the six industries (Figure 2.6). The region is also home to the manufacturing of small boats for local fishing industries or leisure pursuits and some ship repair. Shipbuilding accounts for 16% of the region's employment, reflecting the importance of the cruise tourism industry in the Caribbean. At the same time, there are some important differences at national level among the countries of the region. The Dominican Republic has the largest value added in freight and passenger transport. It also the highest GDP in the region and fa relatively diversified economy based on agriculture, mining, services and tourism. Suriname, the Bahamas, Jamaica, Haiti and Belize have a larger value added in marine fisheries, although this sector is almost certainly dwarfed by tourism in each country.

Many of the island countries are tourism-dependent economies, in particular Antigua and Barbuda, Dominica, Grenada, Grenadines, and Saint Vincent. Only a few Caribbean countries have diversified economies based on commodities, services and other natural resources. An example is Belize, which while highly dependent on tourism, also exports oil since 2005. One important factor for all Caribbean countries is their vulnerability to natural disasters and climate change. The island country of Dominica suffered damages estimated at 226% of its annual GDP during the 2017 hurricane season (World Bank, 2020[7]).

Figure 2.6. Share of value added and employment for six ocean-based industries in selected Caribbean countries, 2015

Total value added in millions of constant 2010 USD and total employment in thousands of persons engaged (2015)

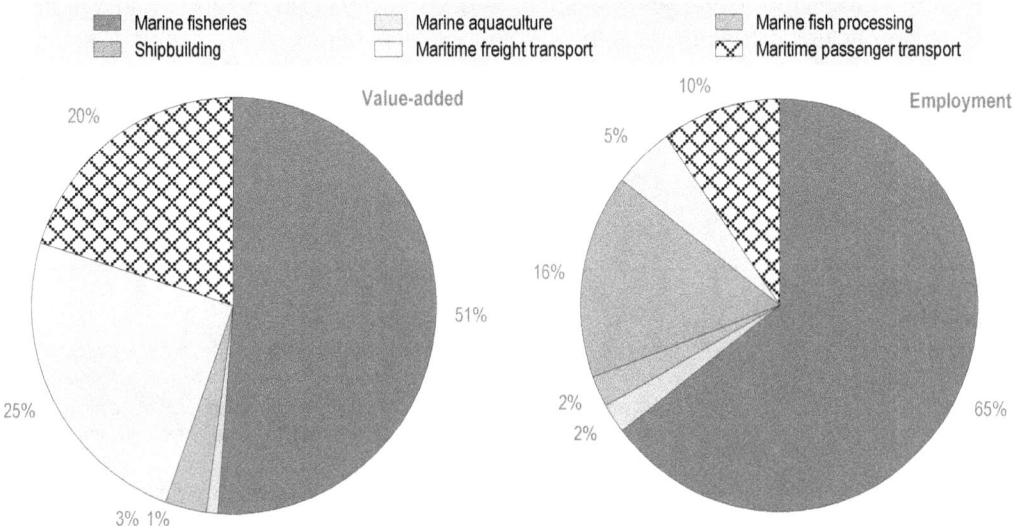

Note: The values presented here are part of ongoing OECD experimental work in building up an ocean satellite account. Future estimates may vary for definitional reasons. The Caribbean countries included are the ones also presented in Figure 2.7.
Source: OECD (2020[4]), Experimental Ocean-Based Industry Database, Directorate for Science, Technology and Innovation.

StatLink https://doi.org/10.1787/888934159240

Figure 2.7. Value added from six ocean-based industries in selected Caribbean countries

Total value added in millions of constant 2010 USD (2015)

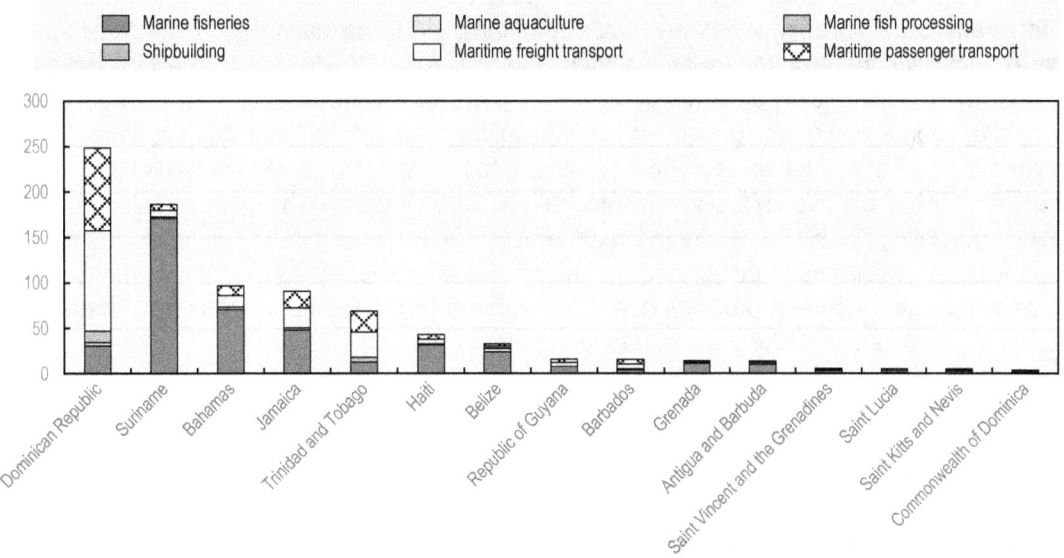

Note: The values presented here are part of ongoing OECD experimental work in building up an ocean satellite account. Future estimates may vary for definitional reasons.
Source: OECD (2020[4]), Experimental Ocean-Based Industry Database, Directorate for Science, Technology and Innovation.

StatLink https://doi.org/10.1787/888934159259

Trends in selected ocean-based industries

This section explores trends across developing countries in several specific ocean-based industries: seafood production (including fishing, aquaculture and fish processing); coastal and marine tourism (including the cruise industry); extractive industries (e.g. oil and gas and seabed mining); transport and logistics industries (including freight, passenger transport and ports); manufacturing, shipbuilding and repair industries; renewable energy; and bio-marine resources. As shown already in the previous section and in Table 2.2, some ocean-based sectors account for a considerable share of GDP in developing countries.

Table 2.2. GDP dependence on selected ocean-based industries

Selected ocean-based industries	Examples
Offshore oil and gas	• In Angola, offshore oil production and its supporting activities contribute about 50% of the country's GDP and around 89% of exports • In Nigeria, the oil and gas sector accounts for about 10% of GDP and around 86% of exports revenue (70% of government revenue), with half the production coming from offshore installations
Tourism	• SIDS are particularly dependent on the tourism sector, with two out of three SIDS relying on tourism for 20% or more of GDP • In Antigua and Barbuda, Belize, Maldives, Saint Lucia, and Fiji, the total contribution of tourism to GDP exceeds 40% • In Cabo Verde, tourism represents more than 50% of GDP • In the Seychelles, tourism represents 65% of GDP
Marine fishing	• Value added of marine fisheries represent up to 6% of GDP for low-income countries and 8% for lower middle-income countries

Seafood production: Fishing, aquaculture and fish processing

Capture fisheries and aquaculture are important sources of protein in human diets throughout the world (OECD/FAO, 2019[8]; FAO, 2020[9]). Small-scale fisheries remain the backbone of socio-economic well-being for many coastal communities and especially for developing countries in the tropics, where the majority of fish-dependent countries are located (Golden et al., 2016[10]). As noted in Table 2.2, value added from marine fisheries represents up to 6% of GDP for low-income countries and 8% of GDP for lower middle-income countries. Beyond providing sustenance for communities, capture fisheries and aquaculture are part of a complex seafood value chain that starts with catching or harvesting raw materials as input (e.g. fish, crustaceans), adding value to the raw materials through various processes, and marketing and selling finished products to customers. There are almost as many value chains as species, with very localised networks in the case of many small-scale fisheries and extensive global networks of supply and trade that connect production with consumers in multinational industrial fisheries (Rosales et al., 2017[11]). Fish is one of the most traded food commodities.

Fishing

According to OECD calculations, the value added from marine fishing alone is highest in the grouping of lower middle-income countries, at approximately USD 40 billion in 2015, followed by high-income countries (approximately USD 20 billion) and upper middle-income countries (USD 10 million) (Figure 2.8). The lower middle-income countries host largely coastal, artisanal fisheries, which represent very modest value added. The totals are still conservative, since approximations in many countries often do not take into account the informal, small-scale fisheries and subsistence sectors, which are often important in many developing countries. The lack of data on these fisheries is problematic for policy making at national and regional levels, with impacts on the effectiveness of development assistance and sustainable fisheries management efforts. Indeed, while it is often believed that large-scale fishing has the biggest impacts on fish stocks the most, in fact, the combined pressure of numerous small-scale and subsistence fishers also matters, especially in tropical countries where the abundance of stocks is relatively low. Collecting economic information on artisanal fisheries contributes to better target support policies, for example to identify those fisheries or fleet segments that may be most vulnerable to economic shocks (Delpeuch and Hutniczak, 2019[12]).

Employment data show the number of marine fishing jobs are highest in lower middle-income countries, with about 12 million people employed in this industry in 2014 (Figure 2.9). Upper middle-income countries have the next highest number of marine fishing jobs, with approximately 5 million people employed, followed by high-income countries (just over 500 000 people employed) and lower middle-income countries. While these are very conservative values, based mainly on industrial fishing and likely omitting some small-scale fisheries and family businesses, they highlight interesting distinctions between the different country income groups. For high-income countries, the rather low jobs numbers may be explained by various trends in the industry. The industrialisation of marine fishing and aquaculture activities has been accompanied by a decline in the number of actors in artisanal fisheries. These trends are not yet affecting lower middle-income countries, which account for much of the global employment in marine fishing and aquaculture. However, compliance with national and regional fishing quotas is on the rise and fishing vessel monitoring technologies to check on compliance are spreading. Some examples related to Indonesia in particular are discussed in this chapter. Employment in this industry in upper middle-income countries is starting to slowly decline.

Figure 2.8. Global value added in marine fishing by country income group, 2005-15

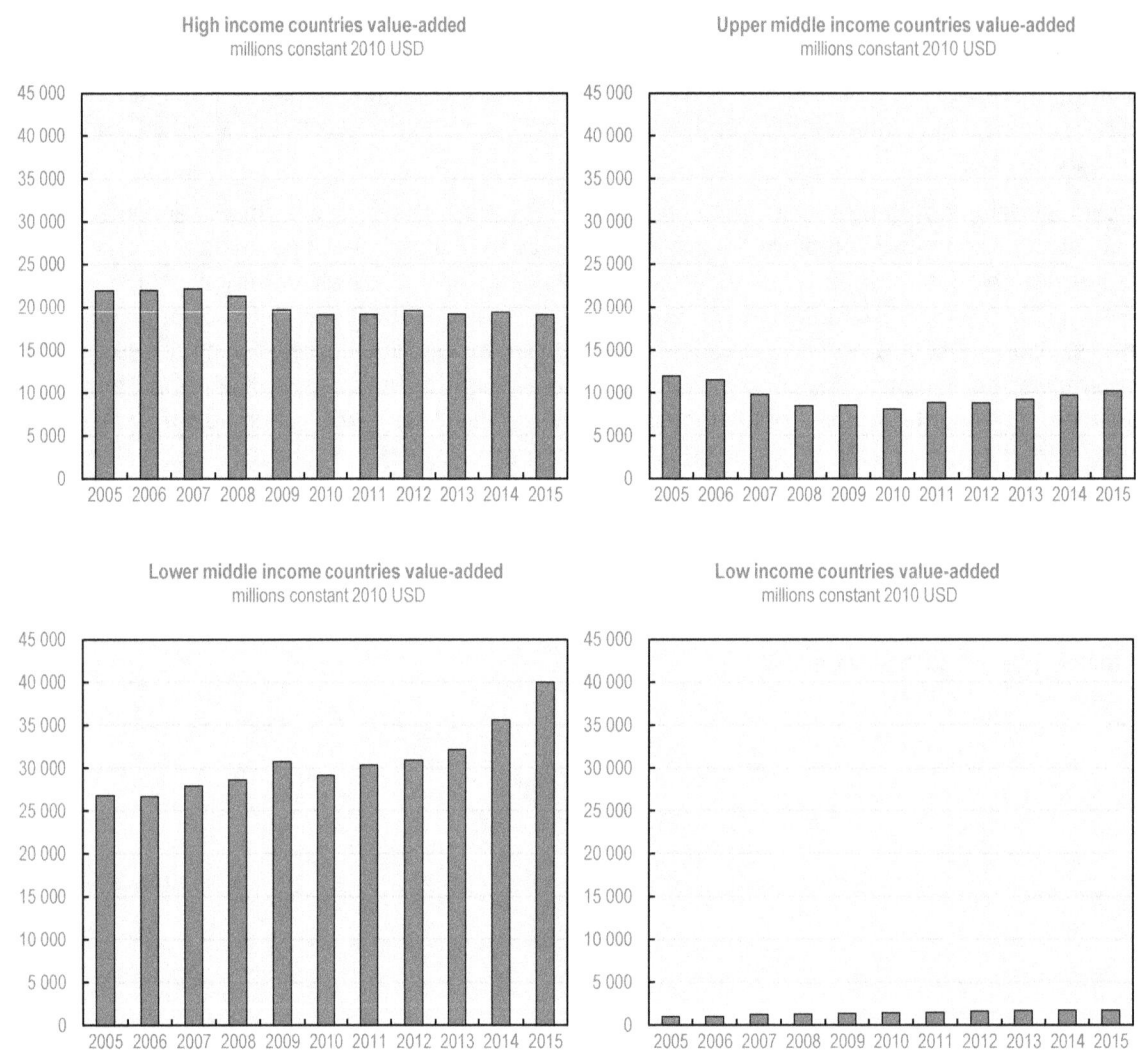

Note: The values presented here are part of ongoing OECD experimental work in building up an ocean satellite account. Future estimates may vary for definitional reasons.
Source: OECD (2020[4]), Experimental Ocean-Based Industry Database, Directorate for Science, Technology and Innovation.

StatLink https://doi.org/10.1787/888934159278

As underscored in this report, policies to support the development of the ocean-based industries in a sustainable manner, combined with necessary marine conservation efforts, also need to be taken up in co-ordination with the local coastal populations to be effective. These issues will be revisited in Chapter 3.

Figure 2.9. Global employment in marine fishing, 2005-15

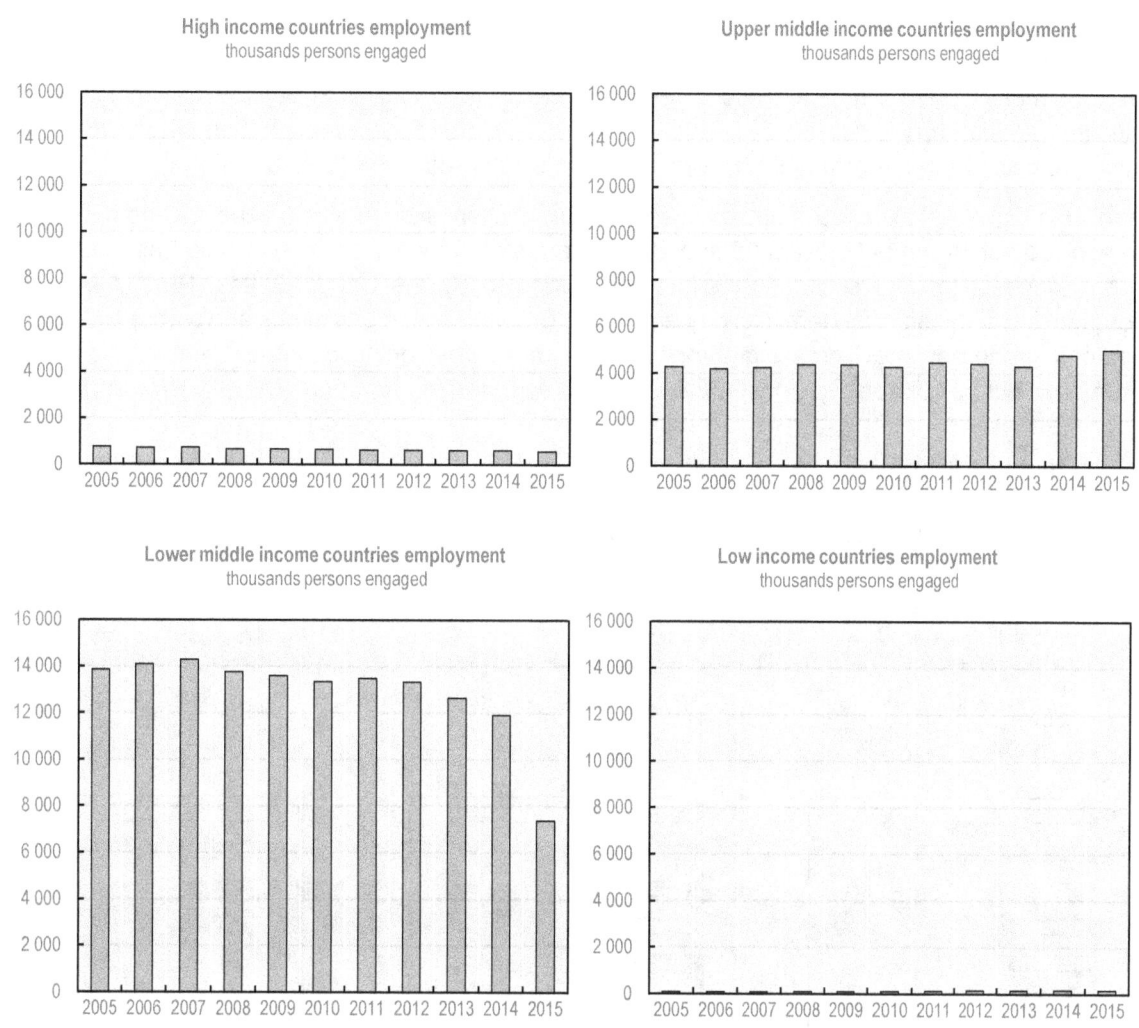

Note: The values presented here are part of ongoing OECD experimental work in building up an ocean satellite account. Future estimates may vary for definitional reasons. The drop in employment in 2015 can be linked to delays in the inclusion of up-to-date production statistics that are part of the model.
Source: OECD (2020[4]), Experimental Ocean-Based Industry Database, Directorate for Science, Technology and Innovation.

StatLink https://doi.org/10.1787/888934159297

Aquaculture

Aquaculture is considered to be one of the sectors with the largest potential for growth (FAO, 2018[13]) and has expanded substantially in recent years, driving up total fish production against a more stagnating trend for wild fish catch. In 2016, global aquaculture production, including both inland and marine production, was 110.2 million tonnes and was worth approximately USD 243.5 billion (FAO, 2018[13]). At least 64.2% of aquaculture production is inland and is dominated by freshwater fin fish such as carp species. Aquaculture in coastal areas includes both species farmed in saltwater ponds, such as shrimp, and species produced in cages and man-made structures either adjacent to or on the coast, such as seaweed and molluscs.

Developing marine aquaculture could be an opportunity for selected developing countries, although it should not come at the expense of coastal ecosystems. Aquaculture can provide an additional source of income for vulnerable coastal populations, who may otherwise rely on farming or fishing. Further, technical improvements in aquaculture systems have greatly increased the feed efficiency of aquaculture in recent years and many systems now achieve a feed conversion ratio similar to poultry systems, albeit with significant variation (Fry et al., 2018[14]). More complex and still at a demonstration stage, open ocean farming projects also have potential for more sustainable fish production (OECD, 2019[15]).

As shown in Figure 2.10 the top countries in seafood production, including both fisheries and aquaculture, have evolved somewhat between 2005 and 2015 in terms of live weight tonnes. China has remained the leader. Indonesia rose to second place in 2015, followed by the United States, Peru, the Russian Federation, India, Japan, Viet Nam, Norway and Chile. Crustaceans (e.g. shrimps and crabs) and molluscs in seafood production are increasingly important for many developing countries, not only for national consumption but as tradeable goods (OECD/FAO, 2019[8]).

Figure 2.10. Top 20 countries in seafood production in live weight tonnes, 2005-15

Total annual production in thousands of live weight tonnes by species division

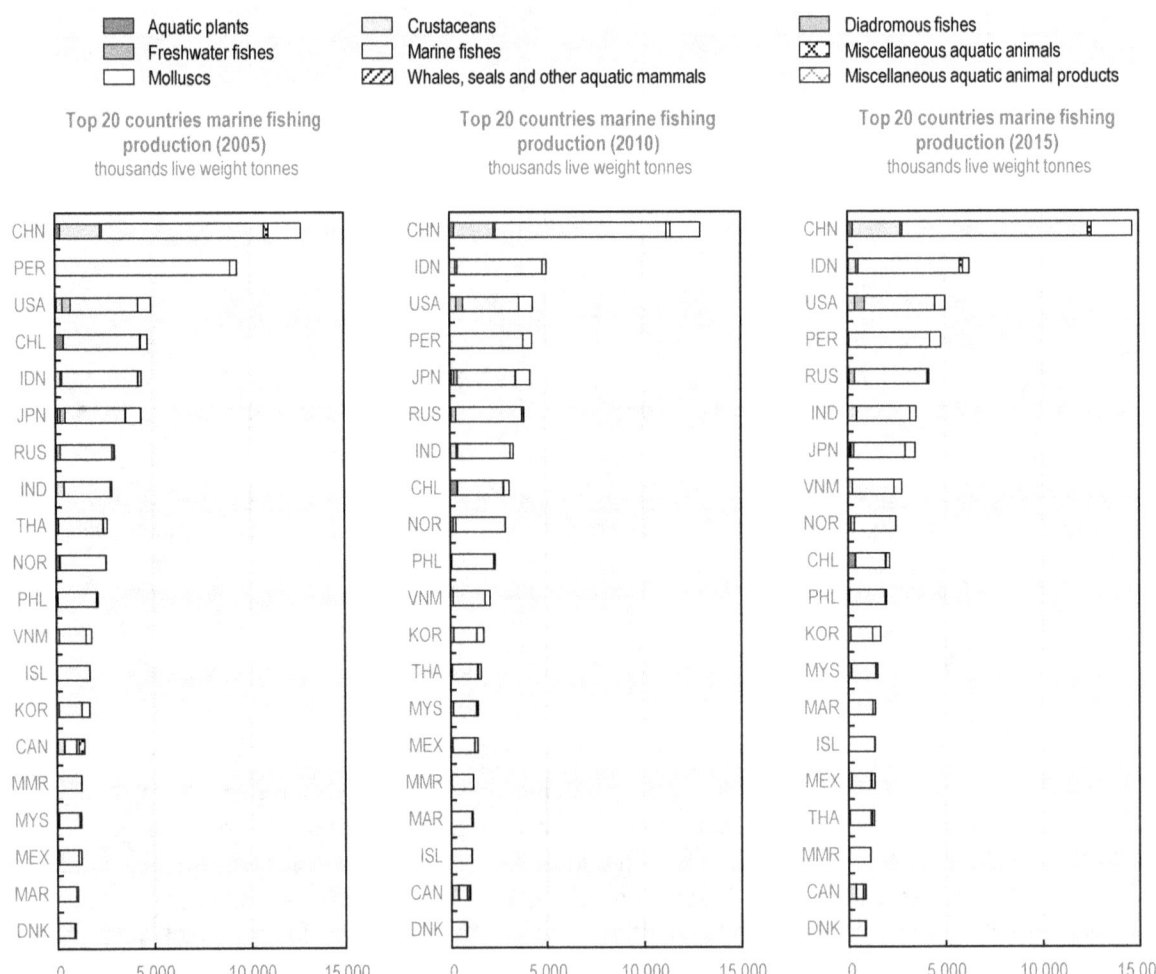

Source: OECD calculations using FAO (2019[16]), Global production by production source 1950-2017, FAO Fishery and Aquaculture Statistics.

StatLink https://doi.org/10.1787/888934159316

Seafood processing

Millions of people and particularly women are involved in artisanal fish processing, making it another important ocean-based industry in developing countries but one that faces some common challenges. Post-harvest facilities such as drying equipment, ice plants and cold storage facilities are often lacking. Such installations are needed for adding value to the seafood product and obtaining better prices, but also to reduce post-harvest losses that occur in artisanal fisheries (Rosales et al., 2017[11]). When no storage facilities are available in the ports with no ice, the fishers sometimes tend to sell their unsold fish at cheaper price or face spoilage of their catches. The Food and Agriculture Organization estimates that approximately 35% of the global harvest is either lost or wasted every year (FAO, 2020[9]). Economic development across the entire fish production system is therefore highly dependent on enhancing post-harvest processing, as well as exploring further sustainable fishing practices (e.g. certifications and eco-labels), as is discussed further in Chapter 3).

Looking ahead, the impacts of overfishing, climate change, coastal pollution, biodiversity loss and illegal, unreported and unregulated fishing will take a toll on seafood production, as they add to the inherent challenges of artisanal fisheries. Some countries will increasingly need more effective strategies for marine conservation and sustainable fisheries management to rebuild stocks for nutritional security (Hicks et al., 2019[17]). The economic downturn provoked by the COVID-19 crisis will particularly impact fish trade[1] and local economies that are reliant on exports.

Coastal and marine tourism including cruise shipping

Tourism is today one of the key sectors in the global economy. In 2019, tourism's direct, indirect and induced impact represented 10.3% of the world's GDP and accounted for approximately 330 million, or one in ten, jobs around the world (World Travel and Tourism Council, 2020[18]). Global tourism has grown significantly, with an estimated 1.5 billion international arrivals in 2019, an increase of 3.8% over the previous ten year increase and well beyond forecasts (World Tourism Organization, 2019[19]). Similarly, global expenditures on travel between 2000 and 2018 more than doubled, from USD 495 billion to USD 1.5 trillion, and now represent 7% of global exports in goods and services. The consumption patterns of the middle classes in OECD and emerging countries are driving the high demand for coastal tourism and cruise tourism, particularly in developing countries (OECD, 2016[1]). The COVID-19 crisis may have lasting effects on the tourism sector, however, with the international tourism economy projected to decline by as much as 70% in 2020 (OECD, 2020[20]).

There are major regional differences in countries' reliance on tourism for their economies. The OECD (2018[21]) estimates that in 2016, tourism represented around 4% of GDP across OECD countries. In contrast, tourism is the main economic sector in several developing countries and an important source of foreign exchange, income and jobs. In Kenya for example, tourism is an important sector, representing 8.8% of GDP in 2018 and attracting up to two million foreign visitors a year, mainly in national parks and along the coast. In some years, coastal tourism can account for some 60% of the revenues. SIDS are also particularly dependent on the tourism sector: two out of three SIDS rely on tourism for 20% or more of their GDP (OECD, 2018[22]). As visitors are frequently concentrated along coastlines, some coastal areas generate up to 80% of total GDP in some countries (Tonazzini et al., 2019[23]). In the Maldives, for example, tourism contributes up to 40% of national GDP. The contribution of direct and indirect tourism to GDP is more than 50% in Cabo Verde, and more than 40% in Antigua and Barbuda, Belize, Saint Lucia, and Fiji It represents 65% of GDP in the Seychelles (Monnereau and Pierre, 2014[24]).

Marine and coastal tourism are highly dependent on the quality of natural ecosystems to attract visitors, as they rely on the recreational value of beaches and clean waters. Yet unmanaged tourism is contributing to ecosystem degradation and fragility, jeopardising it's the sector's own economic sustainability. Climate change vulnerability is also a risk for countries that rely on tourism, seen for instance in coral reefs bleaching events, with the greatest risk in small island states where tourism is also the largest sector of the national economy and the largest employer (Scott, Hall and Gössling, 2019[25]). Other challenges affecting the sector include beach degradation resulting from sand harvesting, mangroves deforestation and an ever-growing coastal population that pressures coastal ecosystems.

Cruise shipping is a major component of the tourism industry and has been growing globally. Small island countries are especially popular destinations for cruise ships Figure 2.11. For example, 825 420 cruise passengers visited Antiqua and Barbuda alone in 2018, a 92% increase since 2000 (Eastern Caribbean Central Bank, 2019[26]).

Figure 2.11. Annual cruise passenger visitors to Caribbean island countries, 2000-18

Millions of passengers

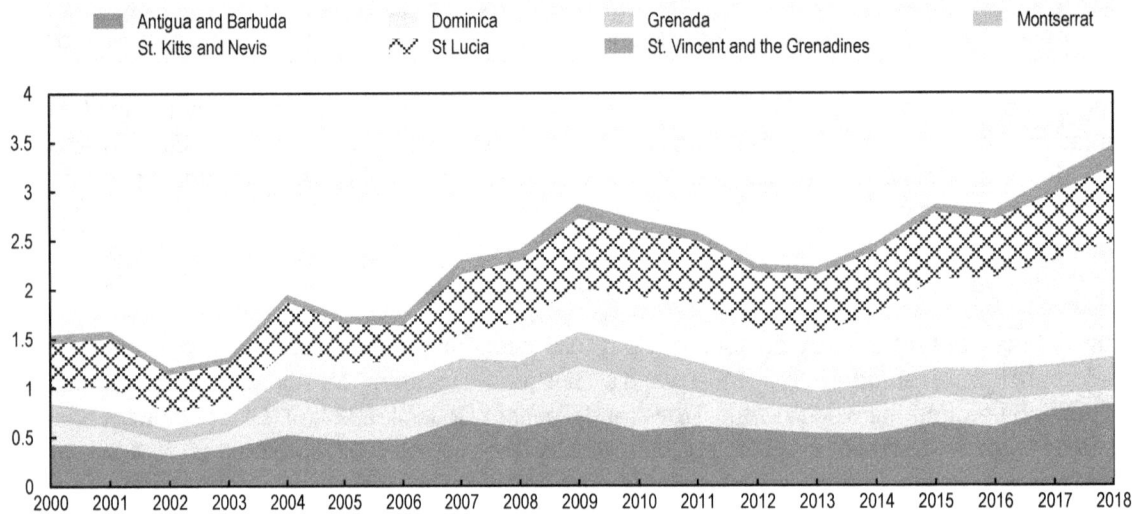

Source: Eastern Caribbean Central Bank (2019[27]), Tourism Annual Report, https://www.eccb-centralbank.org/statistics/tourisms/comparative-report.

StatLink https://doi.org/10.1787/888934159335

As noted elsewhere in this report, an increasingly trenchant issue for many developing countries is balancing the promotion of commercial activities that cater to foreign demand and the need to address environmental concerns. Global tourism is a case in point, as it is has significant adverse environmental impacts by placing pressures on domestic freshwater supplies, food systems and waste disposal systems in particular (OECD, 2018[21]). Chapters 3 and 4 present examples of policy solutions and practices to improve the sustainability of marine and coastal tourism and other ocean-based industries. Another issue is the impacts of the COVID-19 pandemic on the tourism industry and uncertainly over how long they might last. It is difficult to predict how the industry's recovery will play out, given the extent of the economic damage, the reduced purchasing power of many potential travellers and the possibility that tourists may be reluctant to travel, especially to countries without well-developed health systems. Any mid- to long-term governmental support should ideally steer the industry to more sustainable practices, backed by efficient policy and economy instruments.

Extractive industries including oil and gas and seabed mining

Oil and gas

The offshore oil and natural gas industry represents the largest share of today's ocean economy and contributes to many developing economies, particularly in Africa and Latin America, despite important environmental externalities (OECD, 2016[1]). Projects in Indonesia, Malaysia, Myanmar, Thailand and Viet Nam are also ongoing, with around 60% of current production in the Southeast Asia region coming from offshore fields located in shallow waters of less than 450 metres in water depth. Offshore projects generated nearly USD 90 billion of cash flow for publicly traded exploration and production companies in 2019, the third strongest year of the past decade in terms of revenues (Bousso, 2020[28]). The COVID-19 pandemic brought the industry to a sudden halt in early spring 2020 as demand collapsed at a time when supply, already overabundant, was still significantly increasing (IEA, 2020[29]). This will have strong economic impacts on many developing countries.

While Nigeria and Angola are Africa's largest oil and gas producers, an unprecedented number of other African countries – among them Ghana, Mauritania, Mozambique, Senegal, Somalia and South Africa – are extending new exploration licenses to offshore companies (Beckman, 2019[30]). The oil and gas sector accounts for about 10% of Nigeria's GDP and around 86% of its exports revenue, which represents 70% of total government revenue (OPEC, 2020[31]). In Angola, oil production and its supporting activities contribute around 50% of the country's GDP and around 89% of exports. Box 2.1 gives an overview of extractive industries in Africa.

The momentum for new extraction licenses built on the results of recent oil and gas exploration programmes, particularly from the Atlantic coast. These include the discovery in West Africa of large deposits off the coasts of Senegal in the MSGBC basin (Mauritania, Senegal, Gambia, Guinea-Bissau and Equatorial Guinea), all since 2015. Several countries in the region have been working to grow and structure their local industry through training in particular, while also strengthening regulatory institutions to deal especially with the many environmental aspects of the developments. Given the uncertainty surround oil demand and price recoveries in the short to medium term, it is conceivable that investments in some offshore oil and gas projects will be delayed or cancelled due to low prices stemming from the reduced demand and oversupply.

Box 2.1. Extractive industries in Africa

Africa accounts for much of the exponential growth in the world's population, and a quarter of the world's population is expected to be African by 2050. Although Africa holds large natural resources and demonstrates economic dynamism, 46% of the population still live in extreme poverty (UNECA, 2016[32]). In this context, African countries with different connections to the ocean in view of their geography and history are seeking new sources of sustainable economic activities. About 70% of them (38 out of 55 countries) have a coastline including SIDS (Monnereau and Pierre, 2014[24]), and several of these coastal countries are in the process of developing ocean strategies to tap into their marine natural resources. Many of these countries are looking at the potential of ocean-based extractive industries to help manage population increase and tackle poverty.

Currently, extractive industries contribute significantly to public finance. Some African countries' public revenue is largely dependent on these industries, although revenues still tend to not directly benefit the majority of local communities (EITI, 2018[33]). Approximately 30% of all global mineral reserves are located in Africa. The continent's proven reserves of oil and natural gas constitute 8 and 7%, respectively, of the world's stocks. Minerals account for an average of 70% of total African exports and about 28% of GDP. Ghana, already the world's 10th largest producer of gold, also, has a growing offshore oil sector and an increasing number of exploration activities (Arican Natural Resources Center, 2016[34]). Many African countries still face significant challenges in managing revenue from the oil and gas sector and being effective in improving transparency and accountability. Reliable data on how much oil, gas and mining companies produce are important to accurately calculate taxes and royalties that are based on production.

Seabed mining

Rising demand for minerals and metals, alongside the depletion of land-based resources, is stirring growing commercial interest in exploiting resources on the seabed in national waters and the high seas. Marine mining is already underway in various parts of the world and numerous projects are in operation, but these are located almost exclusively on continental shelves not far from the shore. While many developing countries, particularly SIDS, regard seabed mining as a promising opportunity, marine mining could have significant environmental impacts on coastlines and the high seas if it is managed incorrectly (Miller et al., 2018[35]).

To date, ongoing mining projects are targeted for the most part towards extracting diamonds, phosphates and seabed marine sulphide deposits, for example in Namibia and South Africa as well as China, Japan and Korea. On a global scale, these activities are still small and publicly available economic data on the operations are scarce. Nautilus Minerals, one of the world's first seafloor miners, officially declared bankruptcy in late November 2019 as it was trying to develop a deep sea gold, copper and silver project off the coast of the Papua New Guinea. The project suffered from technical issues, financial setbacks and community opposition.

The International Seabed Authority is currently facilitating international negotiations on the drafting of technical and environmental rules as part of a possible International Mining Code for the high seas (International Seabed Authority, 2019[36]). The Authority has thus far signed 30 contracts for exploration, not exploitation, with contractors. The contractors, either government bodies or companies with sponsoring states, have the exclusive right to explore for their specified categories of resources for up to 15 years and to exploit an initial area of up to 150 000 km². Contractors may also apply more than once for extensions of up to an additional five years. Activities are currently taking place in the Clarion-Clipperton Fracture Zone (Pacific Ocean) and the Western Indian Ocean and on the Mid-Atlantic Ridge. As part of each exploration project, the contractors are required to propose a programme for the training of developing country nationals. As such, mineral extraction is now seen by several developing countries as a promising opportunity. As an illustration, the Seychelles and Mauritius have worked regionally to map their seafloor to delimit their continental shelf and extend their sovereignty rights. Chapter 3 describes this collaboration in greater detail.

At this stage, when it moves beyond exclusive economic zones and towards the high seas, mining remains an experimental industry with still-unknown impacts on the marine environment and biodiversity, particularly in areas where knowledge of the ocean floor and deep-water ecosystems is limited. Much science occurs though with the current deep-sea exploration phases. Scientific atlases of the many newly discovered megafauna, meiofauna and macrofauna species found in the abyss are already available on the International Seabed Authority (2020[37]) website, allowing researchers to access wealth of data. However, although deep sea science is progressing, much uncertainty still remains about any future operational seabed mining in the high seas. It could have a wide range of impacts on marine ecosystems that may be potentially widespread and long-lasting, with very slow recovery rates. Some impacts may also be theoretically irreversible for instance disturbance of the benthic community where nodules are removed; plumes impacting the near-surface biota and deep ocean; and deposition of suspended sediment on the sea-floor. In consequence, a few countries have called a precautionary approach and even a moratorium on seabed mining activities. For example, Fiji called on fellow Pacific Islands Forum countries to support a ten-year moratorium from 2020 to 2030 (Doyle, 2019[38]).

The COVID-19 pandemic is affecting short-term demand for minerals and metals, and this also may affect some ongoing plans. Nevertheless, looking forward, the commercial exploitation of resources will remain high on the agenda of many developed and developing countries. As countries continue their negotiations on the international regime for using commercially the high seas and the necessary preservation of biodiversity, future seabed mining operations should remain a key issue to tackle for policy-makers.

Shipping and passenger transport

Both shipping (i.e. maritime freight transport) and passenger transport have been growing steadily. International maritime trade has grown almost every year, with volume hitting an all-time high in 2018 of 11 billion tonnes loaded (UNCTAD, 2019[39]). The top five ship-owning regions –Greece, Japan, China, Singapore, and Hong Kong, China – account for more than 50% of the world's dead weight tonnage. The main sub-sectors in maritime passenger transport also have been registering growth. Globally, ferries transport approximately two billion passengers a year, on a par with air passenger traffic, and the cruise industry carries some 26 million passengers annually (Cruise Market Watch, 2019[40]). In terms of domestic routes, it is often a challenge for developing countries, particularly SIDS and archipelagos, to link different regions via the sea. Indonesia is an example. While its policies target the development of the domestic freight and passenger transport sectors and the servicing of currently uneconomic supply routes, a 2005 regulation prevents international ships from servicing domestic routes and permits international ships to enter only a limited number of designated ports. The government also subsidises a small number of uneconomic goods shipment lines – sea toll routes – through either a state-owned shipping company or through tenders for any remaining capacity needs.

Overall, the maritime transport landscape has changed markedly in recent years, and the COVID-19 crisis is having significant impact (ITF, 2020[41]). Trade growth had been slowing before the pandemic as supply chains and trade patterns became increasingly regionalised, with some possible new gains for coastal lower middle-income countries. In parallel, technology and services have been playing an expanding role in logistics and services, and sustainability issues have been looming larger on the maritime transport industry agenda (ITF, 2019[42]). Additional challenges are likely to arise if the global economic downturn persists.

Shipbuilding

Shipbuilding, like other manufacturing activities, is influenced by a multitude of factors ranging from global trade, energy consumption and prices to changing cargo types and trade patterns, vessel age profiles, scrapping rates and replacement levels. Existing affects shipbuilding capacity also affects developments and has exceeded for some ship requirements for some time, notably due to market-distortive public support in some countries. There has also been a considerable build up of overcapacity in the shipping industry as growth of the global fleet outstripped that of world seaborne trade by a considerable margin (Gourdon, 2019[43]).

High-income and upper middle-income Asian countries are the dominant shipbuilding market leaders, with China, Japan and Korea together accounting for about 90% of global new-build deliveries in tonnage in commercial ships (UNCTAD, 2019[39]). Several European countries (e.g. Denmark, Finland, France, Germany and Italy) are producing highly specialised vessels such as ferries, offshore vessels, and large cruise ships. European countries also account for about a 50% global market share in the marine equipment industry.

In Indonesia, as in other developing countries, the shipbuilding industry is mainly focussed on supplying the domestic market. While the 2012-25 shipbuilding industry development roadmap aims to increase exports of whole ships, the Indonesian industry is constrained by poor access to finance, skilled labour shortages and a tax regime that incentivises the importation of whole ships rather than maritime parts for construction in domestic shipyards. At the end of the value chain, several low-income countries in South Asia (e.g. Bangladesh, India and Pakistan) are involved in the dismantling of the world's ships (Gourdon, 2019[44]).

The COVID-19 pandemic has generated and is expected to generate many effects on the shipbuilding industry and its broader value chains. Ship demand is driven by global maritime activities that have been severely affected by the pandemic. On the supply side, many shipyards experienced production disruptions in 2020 as governments put in place lockdowns and quarantine measures. In addition, ship orders and ship deliveries have been delayed because it has been difficult for ship owners to meet with shipbuilders and conclude deals. The industry will need to adapt to recover.

Renewable energy

Electrification is a major challenge in many developing countries that remain dependent on imported fossil fuel for energy generation. The cost of fossil fuels is a burden on government budgets, business and households, and disproportionally affects people already struggling with poverty. This is especially the case for SIDS, where on average more than 30% of foreign exchange reserves are allocated each year to cover the cost of fossil fuel imports and where retail energy rates are three to seven times higher than in developed economies (OECD, 2018[22]). To lower the cost of energy and transition towards greener, low-emission development pathways, several renewable solutions are being tested thanks to recent innovations in offshore wind farms and solar and geothermal resources. These renewable energy are however often combined with diesel generators to function effectively in developing countries. The share of renewables in meeting global energy demand is expected to increase by a fifth over the period of 2018-23, reaching 12.4% (IEA, 2018[45]).

Offshore wind in particular is a rapidly growing sector (IEA, 2019[46]). It has expanded at an extraordinary rate over the last 20 years or so in developed and emerging countries, from almost zero to a total global capacity of 18 gigawatts (GW) in 2017. The cost of offshore wind generation has dropped progressively, and projections suggest that offshore wind could reach between 15 to 21 GW per year by 2025 to 2030 (GWEC, 2019[47]). This growth is expected not only in OECD countries and China, but also in several developing countries where offshore wind can expand electricity access and increase the share of renewable resources in the energy mix, thus contributing to the commitments made under the Paris Agreement on climate change. But the long-term impacts of large offshore wind farms on the ocean environment itself is slowly starting to be considered too. The offshore wind sector presents some opportunities but also many specific challenges for low-income countries. Technical difficulties can be vast due to geographic characteristics and remoteness, for SIDS in particular. Offshore wind farms still require rather large upfront investments. Chapter 4 presents examples of potential development financing.

Marine renewable energy – wave, current and tidal energy – is also considered an important potential source of power generation for the transition to a low-carbon future (IEA, 2019[46]). However, ocean energy technologies are for the most part still at the demonstration stage, with only a few prototypes moving towards the commercialisation phase. In many cases, the installation of wind turbines on land or as offshore platforms is not possible due to topographical constraints and competition for space with other ocean-based industries, typically coastal tourism. This is particularly the case for many SIDS that are considering these marine renewable energy options.

The use of geothermal resources can be explored for some tropical islands, particularly volcanic islands. Most geothermal technologies generate stable and carbon dioxide (CO_2) emissions-free baseload power. The International Energy Agency, in 2018 forecasts for renewable energy to 2023, projected that growth in geothermal capacity as projects in nearly 30 countries come on line, with 70% of the growth in developing countries and emerging economies (IEA, 2018[45]). The Asia Pacific region (excluding China) has the largest growth (1.9 GW) over the forecast period. This is driven by Indonesia's expansion, which is propelled by its abundant geothermal resource availability and a project pipeline in the construction phase supported by government policies. Kenya, the Philippines and Turkey follow in terms of capacity growth. Critical issues to resolve concern environmental impacts (e.g. seismicity, possible ground and water contamination, and air pollution) and economic and governance aspects (Table 2.3) (Meller et al., 2018[48]). In the French Caribbean department of Martinique, for example, the first large geothermal plant, named Nemo (New Energy for Martinique and Overseas), was launched in 2016 and then stopped due to technical difficulties and environmental risks linked to the use of ammonia in production. Although pre-development risks continue to pose an important barrier to securing financing for geothermal projects, exploration and construction of facilities in Latin American and Caribbean countries are expected to accelerate.

The policy responses to the COVID-19 crisis may potentially lead to further development in renewable energy systems, as investments in the energy infrastructure are re-evaluated in both developed and developing countries (Birol, 2020[49]).

Table 2.3. Public perceptions of deep geothermal energy

	Negative perception	Positive perception
Environment	induced seismicity	contribution to the renewable energy mix and reduced reliance on imported energy sources such as coal, natural gas, and diesel
	water contamination	low land consumption
	air pollution	local usage
	Noise	robustness of energy system
	damage of flora and fauna	reduction of CO_2 emissions
Economy	Damage to infrastructure, financial risks	economic development of regions
Governance	public participation in planning,	public participation in planning,
	responsibility in case of damages	early and transparent information,
	commitment of public institutions	inclusion of public concerns in planning process

Source: Adapted from Meller et al. (2018[48]), "Acceptability of geothermal installations: A geoethical concept for GeoLaB", https://doi.org/10.1016/j.geothermics.2017.07.008

Marine biotechnologies

To date, the potential of marine bio-resources remains largely untapped, although many developing countries have extensive and valuable marine resources such as corals, sponges and fish. As ocean processes become better known, many countries are developing strategies to foster marine biotechnology for future pharmaceutical drug development and cosmetic products for health and well-being as well as for food production using algae, biofuel, etc. (OECD, 2017[50]). Marine bio-resources research is already essential in many industries, for instance in the pharmaceutical sector for the development of new generations of antibiotics. Marine genetic resources could be at the core of new solutions to fight pandemics.

Already, an increasing number of developing countries have integrated this marine bio-resources dimension in their respective ocean economy strategies, among them the Seychelles, and more are doing the same. However, the gap between developed and developing countries on bioprospecting is growing, with ten developed countries accounting for more than 98% of the patents associated with a gene of marine origin (Blasiak et al., 2018[51]). Despite international conventions on the protection of biodiversity (e.g. the Convention on Biological Diversity and the Nagoya Protocol on Access to Genetic Resources and the Fair and Equitable Sharing of Benefits Arising from their Utilization), several multinationals are using patents to acquire genetic resources or traditional knowledge. A single corporation – BASF, headquartered in Germany and the world's largest chemical manufacturer – registered 47% of all patent sequences based on genes of marine origin. The Yeda Research and Development Co. Ltd., the commercial arm of the Weizmann Institute of Science in Israel, registered more than half (56%) of all university patents, more than the combined claims of 77 other universities. At this stage, there are no internationally agreed definitions concerning the crucial marine genetic resources that are still being discovered. However, negotiations are ongoing for access to and benefit sharing of these resources.

As seen earlier, the impacts of the COVID-19 pandemic could have long-lasting effects on ocean-based industries in general. But they could also accelerate developments in specific emerging ocean-based sectors, for example marine biotechnologies for medical applications. For instance, the test being used to diagnose the novel coronavirus COVID-19 –and in other pandemics such as HIV/AIDS and SARS – was developed with the help of an enzyme isolated from a microbe found in marine hydrothermal vents (Hugus, 2020[52]).

Developing countries should consider exploring and potentially engaging in a sustainable way in these activities. An important first step to avoiding irreversible damage to fragile ecosystems may be linking with existing knowledge and innovation networks to form partnerships and base any future activities on scientific evidence. Chapter 3 highlights some collaborative approaches.

Some cross-sectoral perspectives

The objective of most private investment strategies is to achieve the highest possible return within reasonable risks, whether these are financial, technical or reputational risks. The broader environmental and social consequences of business operations have become an important focus for many ocean-based industries.

Since the UN Ocean Conference in New York in 2017, the importance of demonstrating sustainable practices is growing considerably in various ocean-based industries. The UN Global Compact, for instance, has developed a Sustainable Ocean Business Action Platform that convenes dozens of representatives of ocean-based industries to develop common principles and actions to advance progress towards the Sustainable Development Goals (UN Global Compact, 2019[53]). In 2019, the Compact developed Sustainable Ocean Principles that emphasise the responsibility of businesses to take necessary actions to secure a healthy and productive ocean. Although a strong policy and regulatory framework at national level is crucial to ensure sustainability across ocean-based industries (Chapter 3), more private actors are starting to align with the requirement for improved sustainability based on voluntary commitments.

Another cross-sectoral element concerns the importance of global value chains (GVCs) in ocean-based industries. All countries, including developing countries, aim to attract foreign investments and encourage their businesses to enter new markets. The general environment of the ocean economy is quite competitive including among developing countries, as illustrated in the competition to attract cruise ships in the Caribbean. Moreover, this environment is sensitive to even small changes in costs of doing business, production or trade. Many low-income countries thus face even more challenges and may need support to put in place some preconditions for integration into GVCs. The preconditions can include but are not restricted to open trade and investment regimes. Other important factors include the development of infrastructure and the readiness of institutions, supporting local human capital through education and training, improving the business climate, and the availability of capital (OECD, 2015[54]).

Looking forward: Encouraging a balanced mix of sustainable ocean use and conservation

As the COVID-19 pandemic-induced crisis continues through 2020 and beyond, reduced activity in ocean-related sectors such as marine and coastal tourism will have an impact on developing countries' socio-economic fabric. In view of the wider impacts on the world economy, the evidence developed for this report should eventually contribute to building a baseline to review the positioning of some countries in selected ocean-based industries and to plan for recovery and new developments as additional and better economic evidence is developed.

Despite uncertainties, many of the major trends associated with ocean-based industries will continue. For instance, longer-term demand for marine sources of food, energy, minerals and leisure pursuits is still likely to increase as the global population grows. Some untapped opportunities remain for developing countries to benefit from sustainable development pathways. Improving long-term sustainability should remain a core factor in decisions related to ocean-based industries, as policy makers consider strategies to stimulate their economies once it is safe to do so. The development of ocean-based industries should go hand in hand with preserving marine natural assets and ecosystem services.

Political will and public investment in ocean governance tools are needed to encourage and develop more sustainable practices in relation to the marine environment. Many policy instruments exist – ranging from regulatory command-and-control instruments to economic instruments and information and voluntary approaches – to create the right business environment to attract investments, protect the interest of fragile communities and make sure that any development is respectful of environmental laws. In other words, there is much that can be done to ensure that a sustainable ocean economy is sufficiently supported and that its benefits are shared as broadly as possible. Chapter 3 provides an overview of these instruments and practical examples for selected sectors.

References

Arican Natural Resources Center (2016), *Catalyzing Growth and Development Through Effective Natural Resources Management*, African Development Bank Group, Abidjan, Côte d'Ivoire, https://www.afdb.org/fileadmin/uploads/afdb/Documents/Publications/anrc/AfDB_ANRC_BROCHURE_en.pdf. [34]

Beckman, J. (2019), "Sub-Saharan Africa opening up to offshore investors", *Offshore*, https://www.offshore-mag.com/drilling-completion/article/16763971/subsaharan-africa-opening-up-to-offshore-investors. [30]

Birol, F. (2020), "Put clean energy at the heart of stimulus plans to counter the coronavirus crisis", *IEA Commentaries*, https://www.iea.org/commentaries/put-clean-energy-at-the-heart-of-stimulus-plans-to-counter-the-coronavirus-crisis. [49]

Blasiak, R. et al. (2018), "Corporate control and global governance of marine genetic resources", *Science Advances*, Vol. 4/6, p. eaar5237, http://dx.doi.org/10.1126/sciadv.aar5237. [51]

Bousso, R. (2020), *Offshore oil and gas boom to continue*, https://www.reuters.com/article/us-oil-offshore/offshore-oil-and-gas-boom-to-continue-rystad-idUSKBN1ZD1N6. [28]

Cruise Market Watch (2019), *Statistics - 2018 Worldwide Cruise Line Market Share (webpage)*, https://cruisemarketwatch.com/market-share/ (accessed on 8 November 2019). [40]

de Graaf, S. and L. Garibaldi (2014), *The Value of African Fisheries*, Food and Agriculture Organization, Rome, http://www.fao.org/3/a-i3917e.pdf. [6]

Delpeuch, C. and B. Hutniczak (2019), "Encouraging policy change for sustainable and resilient fisheries", *OECD Food, Agriculture and Fisheries Papers*, No. 127, OECD Publishing, Paris, https://dx.doi.org/10.1787/31f15060-en. [12]

Doyle, A. (2019), "Who is in charge of the high seas?", *Financial Times*, https://www.ft.com/content/dcbc6e94-de26-11e9-b8e0-026e07cbe5b4. [38]

Eastern Caribbean Central Bank (2019), *Tourism Annual 2018*. [26]

Eastern Caribbean Central Bank (2019), *Tourism Annual Report*, https://www.eccb-centralbank.org/statistics/tourisms/comparative-report. [27]

EITI (2018), *EITI in Africa*, Extractive Industries Transparency Initiative (EITI) Oslo, https://eiti.org/sites/default/files/documents/eiti_africa_brief_en.pdf. [33]

FAO (2020), *In Brief: The State of World Fisheries and Aquaculture 2020 - Sustainability in Action*, Food and Agriculture Organization (FAO), Rome, http://dx.doi.org/10.4060/ca9231en. [9]

FAO (2019), *FAO Global production by production source 1950-2017*, FAO Fisheries and Aquaculture - Statistics, August 2019., http://www.fao.org/fishery/statistics/en. [16]

FAO (2019), *Fisheries and Aquaculture Country Profiles: The Republic of Angola*, Food and Agriculture Organization (FAO), Rome, http://www.fao.org/fishery/facp/AGO/en#CountrySector-SectorSocioEcoContribution. [5]

FAO (2018), *The State of World Fisheries and Aquaculture 2018*, Food and Agriculture Organization (FAO), Rome, http://www.fao.org/3/I9540EN/i9540en.pdf. [13]

Fry, J. et al. (2018), "Feed conversion efficiency in aquaculture: Do we measure it correctly?", *Environmental Research Letters*, Vol. 13/2, p. 024017, http://dx.doi.org/10.1088/1748-9326/aaa273. [14]

Golden, C. et al. (2016), "Nutrition: Fall in fish catch threatens human health", *Nature*, Vol. 534/7607, pp. 317-320, http://dx.doi.org/10.1038/534317a. [10]

Gourdon, K. (2019), "An analysis of market-distorting factors in shipbuilding: The role of government interventions", *OECD Science, Technology and Industry Policy Papers*, No. 67, OECD Publishing, Paris, https://dx.doi.org/10.1787/b39ade10-en. [43]

Gourdon, K. (2019), "Ship recycling: An overview", *OECD Science, Technology and Industry Policy Papers*, No. 68, https://doi.org/10.1787/397de00c-en. [44]

GWEC (2019), *Global Offshore Wind Report*. [47]

Hicks, C. et al. (2019), "Harnessing global fisheries to tackle micronutrient deficiencies", *Nature*, Vol. 574/7776, pp. 95-98, http://dx.doi.org/10.1038/s41586-019-1592-6. [17]

Hugus, E. (2020), "Finding answers in the ocean: In times of uncertainty, the deep sea provides potential solutions", *Woods Hole Oceanographic Institution*, https://www.whoi.edu/news-insights/content/finding-answers-in-the-ocean/. [52]

IEA (2020), *Oil Market Report April 2020*, International Energy Agency, Paris, https://www.iea.org/topics/oil-market-report. [29]

IEA (2019), *Renewables 2019: Analysis and Forecast from 2019-2024*, International Energy Agency, Paris, https://www.iea.org/renewables2019/ (accessed on 29 October 2019). [46]

IEA (2018), *Renewables 2018: Analysis and Forecasts to 2023*, International Energy Agency, Paris, https://www.iea.org/reports/renewables-2018 (accessed on 22 July 2019). [45]

International Seabed Authority (2020), *Deep-Sea Taxonomic Atlases (webpage*, https://www.isa.org.jm/vc/deep-sea-taxonomic-atlases (accessed on 5 May 2020). [37]

International Seabed Authority (2019), *Exploration Contracts (webpage*, https://www.isa.org.jm/deep-seabed-minerals-contractors (accessed on 22 July 2019). [36]

ITF (2020), *COVID-19 Transportation Brief: How Badly Will the Coronavirus Crisis Hit Global Freight?*, OECD Publishing, Paris, https://www.itf-oecd.org/sites/default/files/global-freight-covid-19.pdf. [41]

ITF (2019), "Maritime subsidies: Do they provide value for money?", *International Transport Forum Policy Papers*, No. 70, OECD Publishing, Paris, https://www.itf-oecd.org/maritime-subsidies-do-they-provide-value-money. [42]

Meller, C. et al. (2018), "Acceptability of geothermal installations: A geoethical concept for GeoLaB", *Geothermics*, Vol. 73, pp. 133-145, http://dx.doi.org/10.1016/J.GEOTHERMICS.2017.07.008. [48]

Miller, K. et al. (2018), "An overview of seabed mining including the current state of development, environmental impacts, and knowledge gaps", *Frontiers in Marine Science*, Vol. 4, p. 418, http://dx.doi.org/10.3389/fmars.2017.00418. [35]

Monnereau, I. and P. Pierre (2014), *Unlocking the Full Poentital of the Blue Economy: Are African Small Island Developing States Ready to Embrace the Opportunities?*, United Nations Economic Commission for Africa, Addis Ababa, http://dx.doi.org/10.13140/RG.2.1.1928.6001. [24]

OECD (2020), *A Blueprint for Improved Measurement of the International Ocean Economy*, OECD Publishing, Paris. [55]

OECD (2020), *Experimental Ocean-Based Industry Database*, Directorate for Science, Technology and Innovation, OECD, Paris. [4]

OECD (2020), "Tourism polic responses to coronavirus (COVID-19)", *Tackling Coronavirus (COVID-19)*, https://read.oecd-ilibrary.org/view/?ref=124_124984-7uf8nm95se&title=Covid-19_Tourism_Policy_Responses. [20]

OECD (2019), "Innovative approaches to measuring the ocean economy", in *Rethinking Innovation for a Sustainable Ocean Economy*, OECD Publishing, Paris, https://dx.doi.org/10.1787/d71e8b4d-en. [2]

OECD (2019), *Rethinking Innovation for a Sustainable Ocean Economy*, OECD Publishing, Paris, https://dx.doi.org/10.1787/9789264311053-en. [15]

OECD (2018), *Making Development Co-operation Work for Small Island Developing States*, OECD Publishing, Paris, https://dx.doi.org/10.1787/9789264287648-en. [22]

OECD (2018), *OECD Tourism Trends and Policies 2018*, OECD Publishing, Paris, https://dx.doi.org/10.1787/tour-2018-en. [21]

OECD (2017), "Marine biotechnology: Definitions, infrastructures and directions for innovation", *OECD Science, Technology and Industry Policy Papers*, No. 43, OECD Publishing, Paris, https://dx.doi.org/10.1787/9d0e6611-en. [50]

OECD (2016), *The Ocean Economy in 2030*, OECD Publishing, Paris, https://dx.doi.org/10.1787/9789264251724-en. [1]

OECD (2015), "Participation of Developing Countries in Global Value Chains: Implications for Trade and Trade-Related Policies", *OECD Trade Policy Papers*, No. 179, OECD Publishing, Paris, https://doi.org/10.1787/5js33lfw0xxn-en. [54]

OECD (forthcoming), *A Blueprint for Improved Measurement of the International Ocean Economy*, OECD Publishing, Paris. [3]

OECD/FAO (2019), *OECD-FAO Agricultural Outlook 2019-2028*, OECD Publishing, Paris/Food and Agriculture Organization of the United Nations, Rome, https://dx.doi.org/10.1787/agr_outlook-2019-en. [8]

OPEC (2020), *About Us: Member Countries*, Organization of the Petroleum Exporting Countries (OPEC), https://www.opec.org/opec_web/en/about_us/25.htm. [31]

Rosales, R. et al. (2017), "Value chain analysis and small-scale fisheries management", *Marine Policy*, Vol. 83, pp. 11-21, http://dx.doi.org/10.1016/j.marpol.2017.05.023. [11]

Scott, D., C. Hall and S. Gössling (2019), "Global tourism vulnerability to climate change", *Annals of Tourism Research*, Vol. 77, pp. 49-61, http://dx.doi.org/10.1016/j.annals.2019.05.007. [25]

Tonazzini, D. et al. (2019), *Blue Tourism: The Transition Towards Sustainable Coastal and Maritime Tourism in World Marine Regions*, Ecounion, Barcelona, http://www.ecounion.eu/wp-content/uploads/2019/06/BLUE-TOURISM-STUDY.pdf. [23]

UN Global Compact (2019), "30 companies and institutional investors commit to take action to secure a healthy and productive ocean", https://www.unglobalcompact.org/news/4492-10-22-2019. [53]

UNCTAD (2019), *Review of Maritime Transport 2019*, United Nations Conference on Trade and Development (UNCTAD), Geneva, https://unctad.org/en/PublicationsLibrary/rmt2019_en.pdf. [39]

UNECA (2016), *Africa's Blue Economy: A Policy Handbook*, United Nations Economic Commission for Africa (UNECA), Addis Ababa, https://www.uneca.org/publications/africas-blue-economy-policy-handbook. [32]

World Bank (2020), *The World Bank in the Caribbean: Overview of the Caribbean (webpage)*, https://www.worldbank.org/en/country/caribbean/overview. [7]

World Tourism Organization (2019), *International Tourism Highlights: 2019 Edition*, http://dx.doi.org/10.18111/9789284421152. [19]

World Travel and Tourism Council (2020), *Economic Impact Reports*, World Travel and Tourism Council (WTTC), https://wttc.org/Research/Economic-Impact. [18]

Notes

[1] For more information on the impacts of COVID-19 on the fisheries sector please see: https://read.oecd-ilibrary.org/view/?ref=133_133642-r9ayjfw55e&title=Fisheries-aquaculture-and-COVID-19-Issues-and-Policy-Responses

3 Policy instruments and finance for developing countries to promote the conservation and sustainable use of the ocean

This chapter examines the regulatory (command-and-control) and economic policy instruments as well as other finance mechanisms available for countries to create incentives and generate revenue for the conservation and sustainable use of the ocean. It focuses on sustainable fisheries and aquaculture, sustainable tourism, and reducing marine pollution, among others. In addition to its discussion of economic instruments, the chapter highlights three other mechanisms – conservation trust funds, blue carbon payments, and private sector concessions and community management approaches – and the role they can play in supporting the transition to a sustainable ocean.

The need for coherent policy approaches to foster sustainable ocean economies

The ocean economy covers many different sectors and each often has its own governing institutions and policy objectives. Policy making for the ocean economy is therefore frequently fragmented, with responsibility split between multiple sectoral-level ministries. The interconnected nature of ocean, seas and marine resources and the economic sectors they support means that more holistic approaches are needed to ensure policy coherence, identify and manage trade-offs between the sectors, and take advantage of synergies where policies can deliver benefits to multiple sectors. The institutional frameworks in place in this context are also important and can help to foster or hinder such approaches.

Several countries have created dedicated ministries with an overarching responsibility for policy making in the ocean economy. Among these are Barbados, Cabo Verde and Indonesia. Cabo Verde, for instance, created a Ministry for the Maritime Economy in 2018 with a broad portfolio with respect to the ocean that includes traditional maritime sectors such as transport as well as fisheries. Indonesia has taken a different approach, creating the Co-ordinating Ministry of Maritime Affairs and Investments in 2015 to oversee and integrate action across a range of other national ministries including Maritime Affairs and Fisheries, Energy and Mineral Resources, and Tourism and Transport. Given that these co-ordination mechanisms are relatively new, it is not yet possible to evaluate how effective these have been in developing coherent and co-ordinated policies.

A number of least developed countries (LDCs) have also established institutional arrangements and policy frameworks to use their blue natural capital more sustainably and as a driver of sustainable development. In 2017, Bangladesh established the Blue Economy Cell, an inter-ministerial platform tasked to develop a road map and co-ordinate initiatives among ministries on the sustainable ocean economy. Other countries, among them Cambodia and Mozambique, have folded ocean economy responsibilities into an existing ministry. Cambodia has also a specific focus on "the blue economy development with sustainability", outlined in its 2013-30 National Strategic Plan on Green Growth (National Council on Green Growth, 2013[1]). Overall, implementing the cross-cutting policies needed for the sustainable ocean economy requires dedicated co-ordinating institutions or mechanisms. However, the most effective of these institutional arrangements are likely to be country-specific, dedicated institutions or mechanisms that are able to implement the cross-cutting policy missions needed for the sustainable ocean economy. Box 3.1 highlights some examples of institutional challenges.

Box 3.1. Examples of institutional challenges

- The governance of marine protected areas (MPAs) is a good example of how institutional fragmentation can lead to implementation challenges. In both Indonesia and Antigua and Barbuda, the governance of MPAs is divided among several different ministries. In Indonesia, MPA governance is split between the Ministry of Environment and Forestry and the Ministry of Marine Affairs and Fisheries. In Antigua and Barbuda, MPA governance is split between the Department of Environment, Ministry of Tourism and the Ministry of Agriculture, Fisheries and Barbuda Affairs. In both countries, the ministries have different powers to impose fees and varying capacity to monitor and enforce regulations. For example, in Antigua and Barbuda, the Ministry of Tourism has the power to collect fees from users of national parks while the Ministry of Agriculture, Fisheries and Barbuda Affairs cannot do so on users of the MPA it manages. This division of responsibilities across different ministries can create confusion among stakeholders and impact MPA effectiveness.
- The waste sector in Indonesia is another example of the challenges created by split governance. Indonesia is the world's second largest producer of marine plastic waste, with adverse implications (Jambeck et al., 2015[2]). In response, the government announced a target to

> reduce marine plastic waste 70% by 2025, responsibility for which is shared by the ministries – of Environment and Forestry and Maritime Affairs and Fisheries. However, as most marine plastics come from ineffective solid waste management, with an estimated 83% of solid waste mismanaged (Jambeck et al., 2015[2]), it is challenging for the two ministries to influence this system as it falls under the responsibility of multiple different ministries and local governments.

The policy instruments available to foster a sustainable ocean economy include regulatory (command-and-control) and economic instruments, as well as information and voluntary approaches. Both policy and finance are needed to ensure that ocean sustainability objectives are effectively achieved. Economic instruments, which provide incentives for sustainable production and consumption, are also able to generate government revenue to support the conservation and sustainable use of the ocean. Other available approaches to help mobilise finance for the conservation and sustainable use of the ocean include conservation trust funds, public-private partnerships and co-management. The discussion in this chapter focuses exclusively on policy instruments for managing economic activity within the exclusive economic zone (EEZ) of countries and not on the ocean economy in areas beyond national jurisdiction.

An overview of regulatory, economic and other instruments for the conservation and sustainable use of the ocean

A broad suite of policy instruments is needed to effectively address pressures on the ocean and ensure that the ocean, seas and marine resources are conserved and managed sustainably. Table 3.1 presents some available types of policy instruments.

Given the multiple and diverse pressures on the ocean stemming from different sectors and economic activities (Chapter 2), taking an integrated approach to policy making is essential. Certain policy instruments, such as marine spatial planning (MSP) and MPAs, can serve to help manage and address pressures from several different sectors of the ocean economy (e.g. fisheries, tourism, infrastructure and shipping). Other policy instruments, such as catch limits on fish and gear restrictions, can be used to help address specific pressures stemming from certain sectors. Understanding how to balance sector-specific instruments with broader-scale approaches is important for creating policy mixes that address sustainability in the ocean economy holistically rather than in each sector in isolation.

Regulatory (command-and-control) approaches set out the legal environment in which ocean industries operate inside national EEZs. They impose standards and best practices on access to or use of natural resources. Economic instruments create continuous incentives for actors in the ocean economy to behave in a more sustainable way by putting a price on environmental externalities, and such instruments can also create revenue for governments that can be used to further support conservation and sustainable use. Information and voluntary approaches are intended to influence the behaviour of producers and consumers, for example by providing information to consumers about the underlying production practices of goods and services derived from the ocean. The "multi-aspect" nature of the environmental challenges facing the ocean implies that a mix of different policy instruments will be required (OECD, 2007[3]).

Table 3.1. Policy instruments to promote the conservation and sustainable use of the ocean

Regulatory (command-and-control) instruments	Economic instruments	Information and voluntary approaches
Marine protected areas	Taxes, charges, user fees (e.g. entrance fees to marine parks)	Certification, eco-labelling (e.g. Marine Stewardship Council)
Marine spatial planning and multi-annual management plans	Rights-based management systems (e.g. individually transferable quotas for fisheries)	Voluntary agreements including public-private partnerships (can include, e.g., voluntary biodiversity offset schemes)
Spatial and temporal fishing closures; bans and standards on fishing gear; limits on number and size of vessels; other restrictions or prohibitions on use (e.g. *CITES)	Subsidies to promote biodiversity and reform of environmentally harmful subsidies	Public awareness campaigns (e.g., on plastic waste) and alternatives to single-use plastics
Catch limits or quotas (output controls)	Payments for ecosystem services (PES)	
Standards (e.g. **MARPOL for ships); bans on dynamite fishing	Biodiversity offsets	
Licenses (e.g. aquaculture)	Non-compliance penalties	
Planning requirements (e.g. environmental impact assessments and strategic environmental assessments)	Fines on damages (e.g. oil spills)	
Ban on single-use plastics (e.g. drinking straws and plastic bags)	Levy on single-use plastics (e.g. plastic bags)	
Extended producer responsibility		

Note: *CITES is the Convention on International Trade in Endangered Species of Wild Fauna and Flora; **MARPOL is the International Convention for the Prevention of Pollution from Ships.
Source: Adapted from OECD (2017[4]), *Marine Protected Areas: Economics, Management and Effective Policy Mixes*. https://doi.org/10.1787/9789264276208-en.

Table 3.2 provides an indication of the extent to which selected instruments are able to target or address the different pressures on ocean and marine ecosystems. Further detail is presented in Annex 3.A.

Table 3.2. Pressures on ocean and marine ecosystems and instruments to address these

Pressures on ocean and marine ecosystems	Examples of instruments					
	Marine spatial planning	*EIA and **SEA	Marine protected areas	***ITQ for fisheries	Pollution abatement measure	Sustainable fish certification, eco-labelling
Overfishing	2	1	2	2	0	2
Pollution	2	2	1	0	2	1
Habitat destruction	2	2	2	0	1	1
Climate change	1	1	1	0	2	0
Invasive alien species	1	1	0	0	1	0

Note: *EIA is an environmental impact assessment; SEA is strategic environmental assessment; ***ITQ is an individual transferable quota.
In this table, 0 implies the instrument is unable to address this pressure; 1 implies it has potential to help address this pressure (depending on instrument and context); and 2 implies it has significant potential to address pressure.
Source: Adapted from OECD (2017[4]), *Marine Protected Areas: Economics, Management and Effective Policy Mixes*. https://doi.org/10.1787/9789264276208-en.

Regulatory (command-and-control) instruments

Some of the regulatory instruments available to promote the conservation and sustainable use of the ocean, for instance MSP, are cross-sectoral and can thus have wide-reaching impacts on the ocean economy. MSP is the process by which countries allocate temporal and spatial space to economic activities and environmental protection in their ocean areas. Marine spatial planning can be a resources-intensive as well as time-intensive process as it requires the active involvement of many stakeholders. However, its purpose is to provide a comprehensive planning system that helps to manage competing demands for the ocean and allocate resources in an effective way.

Through the MSP process, countries can ensure that economic, social and environmental objectives are balanced, policy priorities are aligned, and all relevant stakeholders are adequately consulted. As such, MSP can help to underpin the development of a sustainable ocean economy. A best practice guide for MSP, produced by UNESCO and the Intergovernmental Oceanographic Commission (IOC), outlines ten steps for an effective MSP process (Ehler and Douvere, 2009[5]). A key component of MSP is the collection and analysis of data on the current and future condition of ocean areas.

A first step is to build the knowledge base on the marine environment in national waters. Effective MSP requires good knowledge of the geophysical characteristics of the coastlines, solid bathymetric studies (i.e. submarine topography), and the mapping and assessment of marine biodiversity (fauna and flora) and other resources. The larger the area to map, the more complex the exercise. Nevertheless, the benefits are such that some developing countries are more advanced than many developed countries in establishing their MSP, as is the case for Gabon and the Seychelles.

In the Indian Ocean, the Seychelles is in the final phase of developing its MSP, which covers all 1.35 million km², to expand marine protections to 30% of its territory, address climate change adaptation, and support its national ocean economy (Chapter 4). The Seychelles held stakeholder consultations in 2014-15 when it launched a marine spatial plan initiative, in line with the principles set out by the International Oceanographic Commission (Table 3.3). The Seychelles principles were adapted from the 2009 UNESCO-IOC ten-step MSP guide (Ehler and Douvere, 2009[5]). Since then, a number of international research expeditions have contributed to map and assess the marine biodiversity and resources of the Seychelles, among them the National Geographic Pristine Seas mission in 2016 and the Nekton Expedition in 2019.

Its MSP initiative is funded through the Seychelles debt-for-nature swap for the ocean – the world's first – that restructures parts of its national debt into long-term marine conservation funding. Designed by The Nature Conservancy, the Programme Coordination Unit of the United Nations Development Programme (UNDP) and the Global Environmental Facility on behalf of the government of the Seychelles, the agreement is funded by public and private donors (The Nature Conservancy, 2019[6]). A portion of the debt repayments are to be used to fund innovative marine protection and climate adaptation projects through the Seychelles Conservation and Climate Adaptation Trust (SeyCCAT, 2019[7]).

Table 3.3. Marine spatial planning principles used by the Seychelles

Integrated	Address the interrelationship of issues and sectors and of nature and development, as integration can help to create complementary and mutually reinforcing decisions and actions.
Ecosystem-based	Safeguard ecosystem processes, resilience and connectedness, recognising that ecosystems are dynamic, changing and sometimes poorly understood and therefore require precautionary decision making.
Public trust	Marine resources are part of the public domain, not owned exclusively by and not exclusively for the benefit of any one group, so decisions should be made in the interest of the whole community and not any one group or private interest.
Sustainability	Decision making should take into account environmental, economic, social and cultural values in meeting the needs of the present without compromising the ability of future generations to meet their needs.
Transparency	The processes used to make decisions should be easily understood by the public and allow citizens to see how decisions are made, how resources have been allocated and how decisions have been reached that affect their lives.
Participatory	Communities, persons and interests affected by marine resource or activity management should have an opportunity to

	participate in the formulation of ocean management decisions.
Precautionary	Article 15 of the Rio Declaration on Sustainable Development states, "In order to protect the environment, the precautionary approach shall be widely applied by States according to their capabilities. Where there are threats of serious or irreversible damage, lack of full scientific certainty shall not be used as a reason for postponing cost-effective measures to prevent environmental degradation
Adaptive	MSP is a continuing, iterative process that learns and adapts over time.

Source: Ehler and Douvere (2009[5]), *Marine Spatial Planning: A Step-by-Step Approach Toward Ecosystem-based Management*, http://dx.doi.org/10.25607/OBP-43.

One of the challenges of putting effective MSPs in place is the underlying data needs. Greater investment in the collection and analysis of data on the ocean environment (e.g. benthic habitat mapping) is needed for effective MSP and the development of a sustainable ocean economy. (Also pertinent is the role of innovation, discussed elsewhere in this chapter, and the discussion in Chapter 5 of the role of development co-operation.) It is important to note that in addition to MSP, there are various models to achieve integrated ocean management, such as sectoral plans and requirements for impact assessments, and co-existence among different ocean industries.

MPAs are the conservation instrument used most frequently by countries, and they are growing in popularity, expanding to cover 17.3% of national EEZs in 2019 from just 2.1% in 2000 (OECD, 2019[8]). In 2014, for example, the government of Gabon announced the initial creation of the country's MPA network, the largest such network in Africa and home to a wide array of threatened marine life including the largest breeding populations of leatherback and olive ridley sea turtles as well as 20 species of dolphins and whales. As part of its national *Gabon Bleu* (Blue Gabon) initiative, the government is building on the findings of several scientific exploration campaigns, such as the National Geographic Pristine Seas project, which surveyed Gabon's 885-kilometer coastline in 2012 (National Geographic Society, 2020[9]). The network has since been extended several times and should reach 30% of the country's Territorial Sea and Exclusive Economic Zone by 2020. It has been effective in curbing illegal fishing and providing benefits to species and coastal communities in collaboration with the Gabonese national parks agency, the Gabonese Navy, Gabon's national fisheries agency, and coastal communities and companies involved in offshore oil production (U.S. Fish and Wildlife Service, 2017[10]).

The value of ocean ecosystem services is estimated at USD 49.7 trillion per year (Costanza et al., 2014[11]). Yet many MPAs lack sufficient funding to effectively protect the marine ecosystems they contain, for instance by ensuring adequate resources for compliance and enforcement (Gill et al., 2017[12]). Further effort is needed to scale up financing for MPAs and ensure they are effectively managed. As noted in this chapter, economic instruments can help. In addition, the OECD (2017[4]) report, *Marine Protected Areas. Economics, Management and Effective Policy Mixes*, discusses this issue in detail.

Other regulatory instruments include the more traditional standards on fishing gear, quotas on fish catch, commercial fishing permits, emission standards for waterway engines, and fuel sulphur limits for vessels, among many others. Habitat conservation bycatch limits (or individual habitat quotas) also exist, although these are not yet common. Planning tools such as environmental impact assessments (EIAs) and strategic environmental assessments (SEAs) are also used. EIAs can be required to assess the impacts of projects such as offshore windfarms, harbour expansion and dredging, marine aquaculture, and oil platforms and rigs. SEAs tend to be undertaken for larger activities, for example to inform a country's strategy for the development of marine energy (OECD, 2017[4]).

Economic instruments

Economic instruments can play a key role in incentivising sustainability in the ocean economy. Such instruments put a price on an environmentally damaging activity and are able to achieve a given environmental objective at lower cost than more traditional command-and-control approaches. Many economic instruments such as fees, charges and taxes can also generate revenue for governments. If earmarked, the revenue can also be channelled and used for the conservation and sustainable use of the ocean. The different kinds of economic instruments are briefly described in Box 3.2.

> **Box 3.2. Economic instruments that incentivise sustainable use of the ocean**
>
> - **Taxes** Based on the polluter pays principle, taxes place an additional cost on the use of the natural resource or the emission of a pollutant to reflect the negative environmental externalities these generate. As such, taxes create incentives for both producers and consumers to behave in more environmentally sustainable ways. Examples include taxes on dumping waste and on other pollutants at sea.
> - **Fees and charges** A fee or charge is a requited payment to a general government. That is, the payer of the charge receives something in return that is more or less in proportion to that charge. Charges and fees can be used to control access to resources by pricing extraction, for example through a fishing licence fee or entry fee to an MPA.
> - **Tradable permits systems** Under tradable permits systems, rights to harvest or access certain resources are allocated through permits that are limited in number and can be traded between permit holders. In some cases, these permits are associated with a specific quota for the level of resource that can be extracted, as in the case of individual tradable quotas for fisheries. These systems can generate revenue for governments if the permits are auctioned.
> - **Subsidies** Governments pay subsidies to producers to support the production of certain goods or services. Subsidies can have a negative environmental impact if they increase the level of an environmentally damaging activity, for example sand extraction for construction. Conversely, environmentally motivated subsidies are intended to have a positive impact by lowering the cost of economic activities that have lower environmental impacts.
> - **Payments for ecosystem services** Based on the beneficiary pays principle, payments for ecosystems services (PES) are voluntary transactions between service users and service providers that are conditional on agreed rules of natural resource management for generating off-site services (Wunder, 2015[13]). An example of PES for the ocean economy is a scheme that pays for the restoration of mangroves to store carbon (e.g. blue carbon payments) and enhance coastal flood protection.
> - **Biodiversity offsets** Based on the mitigation hierarchy, offsetting is a process whereby unavoidable impacts of development are compensated for the creation of new habitat that is equivalent to areas destroyed. At a minimum, biodiversity offsets aim for no net loss of biodiversity and can provide a net gain where more habitat is created than is lost. Biodiversity offsets can be used in the ocean economy to compensate for the development of infrastructure relating to tourism, ports and natural resource extraction. The OECD (2016[14]) report, *Biodiversity Offsets: Effective Design and Implementation*, elaborates such offsets in greater detail.

More than 110 countries are contributing data to the OECD Policy Instruments for the Environment (PINE) database, which includes information on more than 3 500 environment-related policy instruments. These include economic instruments that are relevant to the conservation and sustainable use of the ocean. Analysis of the PINE data finds 57 countries have implemented 209 ocean-relevant economic instruments, of which 188 are currently in force. The most commonly used instrument is taxes, accounting for just over half of the ocean-relevant instruments in the PINE database, followed by fees and charges (Table 3.4.) Taxes relevant to ocean sustainability generated at least USD 4 billion in 2018, with taxes on ocean-related pollution, transport and energy generating the most revenue (OECD, 2020[15]). Figure 3.1 shows the steady growth in ocean-relevant environmental policy instruments, from 57 in 1990 to 188 in 2020. Examples of a number of these ocean-related economic instruments in developing countries are described below.

Table 3.4. Active ocean-relevant economic instruments in the OECD PINE database

	Number of instruments	Percentage
Environmentally motivated subsidy	18	9.5%
Fees and charges	45	23.9%
Taxes	94	51.6%
Tradable permits system	26	13.8%

Source: OECD (2020[15]), *Policy Instruments for the Environment database*, oe.cd/pine, accessed June 23, 2020.

StatLink https://doi.org/10.1787/888934159354

Figure 3.1. Growth in ocean-relevant environmental policy instruments

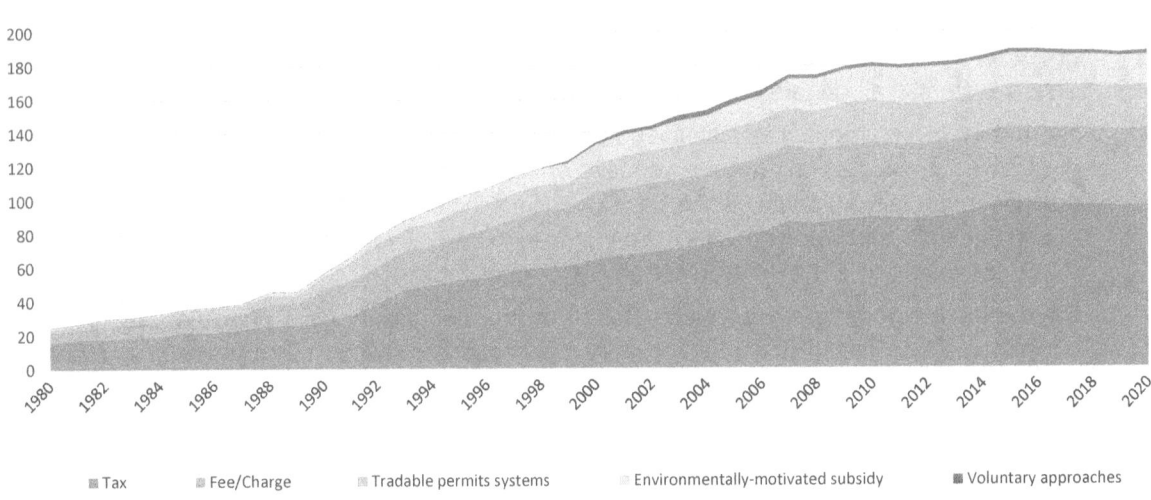

Source: OECD (2020[15]), *Policy Instruments for the Environment database*, oe.cd/pine, accessed June 23, 2020.

StatLink https://doi.org/10.1787/888934159373

Information and voluntary approaches

Information instruments aim to address information asymmetries that often exist among business, government and society. Eco-labels and certification are instruments that have been fairly widely adopted, for example in the case of fisheries. There have been 224 fisheries independently certified as meeting the Marine Stewardship Council standard for sustainable fishing and an additional 94 currently are undergoing assessment (MSC, 2014[16]). Friend of the Sea, among others, is also an important, frequently used certification scheme (OECD, 2011[17]). Examples of other voluntary instruments include negotiated agreements between government and fishers to establish voluntary marine conservation areas.

Policy incentives and finance to conserve the ocean, foster sustainable fisheries, aquaculture and tourism, and manage pollution

Policy incentives and finance to conserve the ocean

Preventing and reversing the loss and degradation of ocean and marine ecosystems are essential to a sustainable ocean economy, as several economic sectors depend on these ecosystems. The fisheries and tourism sectors derive significant value, directly and indirectly, from marine ecosystems. Wildlife tourism, for instance, contributed approximately USD 120 billion to global gross domestic product (GDP) (World Travel & Tourism Council, 2019[18]). Global tourism has increased significantly over the last decade, with an estimated 1.5 billion international arrivals in 2019 (World Tourism Organization, 2019[19]). The continued growth of sectors that derive value from marine ecosystems depends on the ability of governments to manage the ocean and marine resources sustainably. Policy instruments such as marine protected areas can benefit the fisheries and aquaculture sectors, the tourism industry and many other sectors. This section highlights some key economic instruments and finance mechanisms that are available to developing countries to both safeguard the ocean and marine resources and generate revenue for the conservation and sustainable use of the ocean.

Fees and charges for marine protected areas

Many protected areas suffer from chronic underfunding (Watson et al., 2014[20]) (Gill et al., 2017[12]), impacting their ability to protect biodiversity and provide ecosystem services. Marine protected areas tend to have lower establishment costs than do terrestrial areas, but they have higher management costs associated with the logistics of monitoring and enforcing regulations (Bohorquez, Dvarskas and Pikitch, 2019[21]). Given the direct correlation between funding levels and the effectiveness of conservation and sustainable use of biodiversity (Waldron et al., 2017[22]), policies to address MPA funding shortfalls are a priority. User fees have proven to be an effective instrument not only to generate revenue to help cover the cost of managing MPAs but also for controlling access to high-value marine areas (OECD, 2017[4]).

MPA user fees have raised needed revenue in several areas, but they cover the full management costs in only a handful of high-value ecosystems and often must be supplemented by other revenue sources even in popular areas. For example, while the Galapagos National Park in Ecuador generated approximately USD 11.4 million in entrance fees in 2011, the fees did not cover its USD 14.4 million operating cost in that year. Such shortfalls can be covered by other mechanisms such as concessions, as discussed in the subsection on private sector tourism-related concessions and in Thompson et al. (2014[23]). Integrating user fees into a broader financing strategy, for example from government and private-sector concessions, will be important to support many MPAs.

Table 3.5. Marine protected area fees in Kenya

Entry fees for marine protected areas, USD

	Citizen		Residents		Non-residents	
	Adult	Child	Adult	Child	Adult	Child
Entry fee for marine parks						
Kisite Mpunguti	2.10*	1.22*	2.94*	1.66*	17	13
Molindi/Watamu/Mombasa/Kiunga	1.27*	1.22*	2.94*	1.66*	17	13

Boats	All
Fee per day	2.94
Annual pass (private)	50.58
Annual pass (commercial)	147.55

Note: Prices for citizens and residents are charged in Kenyan shillings (KSH); prices for non-residents are charged in US dollars. For ease of comparison, resident fees shown in this table are converted to USD at the exchange rate in December 2019 of USD 1 = KSH 102.2
Source: Kenya Wildlife Service (2019[24]), *Conservation Fees*, www.kws.go.ke/sites/default/files/parksresorces%3A/KWS%20Conservation%20Fees%20Poster.pdf.

StatLink https://doi.org/10.1787/888934159392

Box 3.3. Protected area fees: The example of Raja Ampat, Indonesia

- The marine protected area in the Raja Ampat archipelago of Indonesia offers an example of good practice in using fees. In this MPA, located in the province of West Papua, a longstanding collaboration between environmental non-governmental organisations (NGOs), the local government and the community created a Regional Public Service Agency (BLUD). The BLUD is authorised to collect an entrance or green fee to pay for management of the marine park. The green fee is IDR 1 million (Indonesian rupiah) for foreigners and IDR 500 000 for Indonesians. In addition to funding management of the park, the green fee revenue has been used to engage with the local community on projects through the community welfare fund. This fund allocates at least IDR 1.5 billion (the equivalent of approximately EUR 100 000) annually to villages in and adjacent to the conservation area for economic, social and environmental development programmes. This has improved, relations between the park authorities and local communities, and local fishers now help to enforce the park regulations.

- However, Raja Ampat remains the first and, as of 2019, only area in Indonesia to use the BLUD instrument, in part due to the significant regulatory hurdles to creating a BLUD. One of these is the need to first establish a technical management authority, a process that involves several separate pieces of legislation at different levels of government and, in Raja Ampat, required the strong leadership of the local authorities in collaboration with NGOs. Scaling up this kind of approach, particularly where an MPA contains ecosystems of high value to tourists, can add to the resources available for the conservation and sustainable use of marine ecosystems. However, local stakeholders from governments, communities and businesses are likely unaware of these approaches or unable to operationalise them successfully. While the Raja Ampat example shows that such approaches can work, fewer regulatory hurdles would no doubt facilitate their uptake.

Source: KKP (2016[25]) *Tentang UPTD KKP Raja Ampat* (About Raja Ampat's UPTD KKP), http://www.kkpr4.net/index.php?page=page&id=32

There are challenges and opportunities in the use of fees and charges. Lack of defined access points to marine areas can make it difficult to implement user fees in MPAs. This creates problems of enforcement, as generally MPAs can be accessed on all sides and the cost of patrols can be high (OECD, 2017[4]). Social factors can also reduce the effectiveness of MPA fees, for example if they lead to the exclusion of local populations from traditional use areas. This can lead to conflict between local communities and the MPA, undermining enforcement of the protected area and thus its ecological impact (Pollnac et al., 2010[26]). Many countries use a two-tiered fee system and charge foreign visitors higher fees, as is the case of entrance fees to MPAs in Kenya (Table 3.5). In some areas, tourists pay more than residents. Differentiated fees can enable developing countries to generate revenue from tourists who visit from wealthier, more developed regions while allowing affordable access to local users. Finally, institutional factors can also hamper or even prevent the implementation of user fees. In Antigua and Barbuda, for example, three government agencies share responsibility for managing MPAs. One of these, the Ministry of Agriculture, Lands, Fisheries and Barbuda Affairs, does not have the authority to collect fees in the MPA it manages, while the other two government bodies, the Ministry of Tourism and the Department of Environment, have such authority. In consequence, some MPAs collect access fees and others do not. Further, the fees collected are often returned to the central budget, weakening the incentive for managers to collect them. Indonesia offers an illustration of how these problems can be overcome through a combination of bespoke institutions and benefit sharing with the local community (Box 3.3) (OECD, 2017[4]).

Conservation trust funds

Given the need for broad finance strategies for safeguarding ocean resources, mechanisms that can support a range of interventions are important. Conservation trust funds (CTFs) are independent and private legal entities that can both generate finance and support efforts for the conservation and sustainable use of the ocean. CTFs are used to fund a broad range of projects including the management of protected areas, Payments for Ecosystem Services (PES) programmes and other sustainable development activities. CTFs can therefore play a key role in ensuring sufficient resources are directed towards the conservation and sustainable use of marine natural capital, helping to ensure sustainability in the ocean economy.

Similar to other trust funds, CTFs often rely on long-term funding such as an endowment, at least initially. However, CTFs can also diversify funding mechanisms to include other sources, allowing them to blend finance from public and private sources such as taxes and grants. Not only are CTFs flexible in terms of the revenue they can receive. They can also distribute the money through a variety of mechanisms such as grants, loans and payments for ecosystem services (Figure 3.2). In this way, they act as a financing mechanism and also can catalyse the growth of financing mechanisms such as blue carbon projects.

Figure 3.2. Potential structure of a conservation trust fund

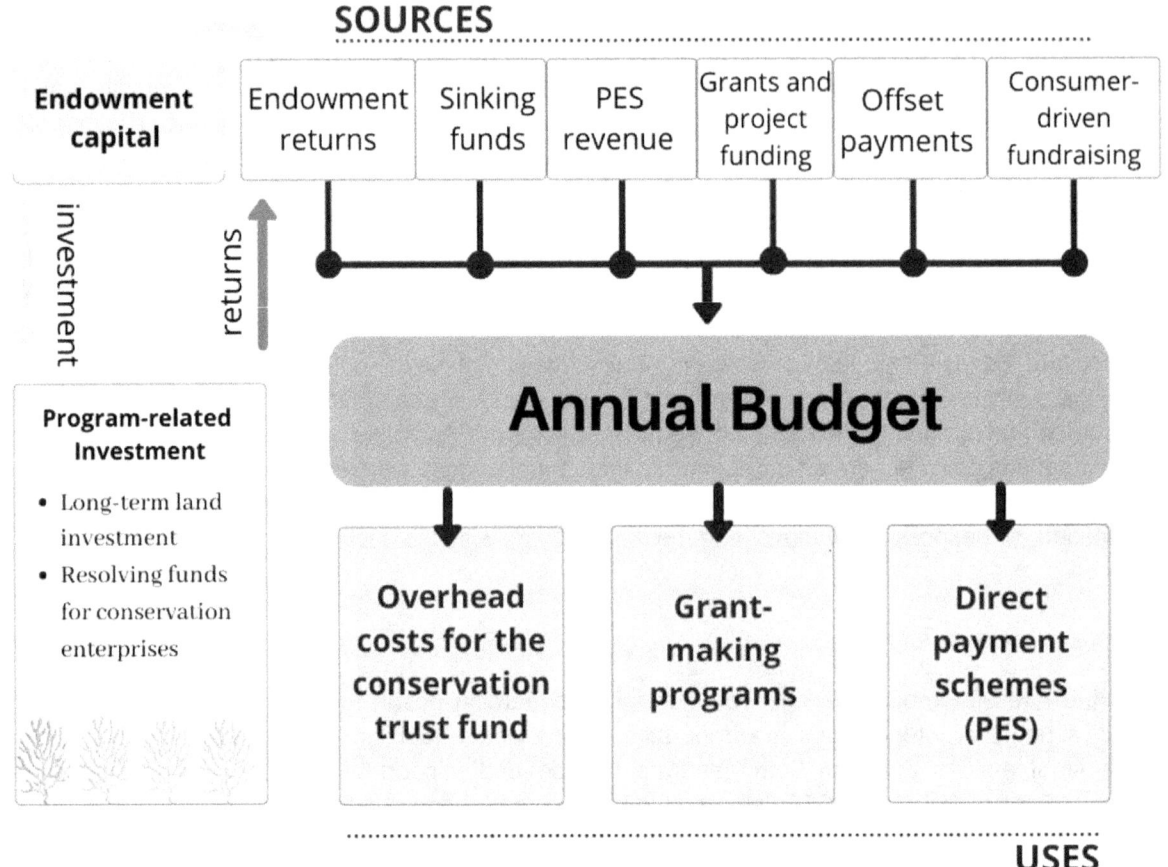

Source: Adapted from Iyer et al. (2018[27]), Finance Tools for Coral Reef Conservation: A Guide, Conservation Finance Alliance

CTFs are a popular tool for raising and distributing finance for the conservation and sustainable use of terrestrial and marine biodiversity. More than 80 are either in development or operational globally. At least 6 of the funds in operation focus specifically on marine and coastal conservation. Antigua and Barbuda, Indonesia, and Kenya are using or setting up CTFs to raise and distribute finance for the conservation and sustainable use of marine biodiversity (Table 3.6). The Kenyan trust fund was mandated by the Wildlife Act of 2013 and will primarily focus on the management and restoration of national parks.

Table 3.6. Overview of select conservation trust funds

Country	Name	Operational?	Year founded	Target size	Funding source	Notes
Antigua and Barbuda	Marine Ecosystems and Protected Area Trust	No	2015	n.a.	Regional Caribbean Biodiversity Fund	Has not yet received an endowment, but has administered funds from the Caribbean Biodiversity Fund
Indonesia	Blue Abadi	Yes	2016	USD 38 million	Endowment	Protected area and community development

Kenya	Kenya conservation trust fund	No	2017	NA	NA	Kenya Wildlife Service and Conservation International are in the process of creating it

Source: Caribbean Biodiversity Fund (n.d.[28]), *Antigua and Barbuda: The Marine and Ecosystem Protected Area, Trust, Inc.* https://www.caribbeanbiodiversityfund.org/antigua-and-barbuda; MPA News (2017[29]), *Financing Spotlight: Blue Abadi, a $38-million Trust Fund to Support MPAs in the Bird's Head Region of Indonesia*, https://mpanews.openchannels.org/news/mpa-news/financing-spotlight-blue-abadi-38-million-trust-fund-support-mpas-bird%E2%80%99s-head-region; Kenya Ministry of Environment and Forestry (n.d.[30]), *Independent Trust Fund is Key to Securing Wildlife Assets, PS*, http://www.environment.go.ke/?p=3393.

The Marine and Ecosystems and Protected Area Trust (MEPA) in Antigua and Barbuda is affiliated to the Caribbean Biodiversity Fund (CBF). The CBF was created to address the second of the two goals of the Caribbean Challenge Initiative:[1] "to have in place fully functioning sustainable finance mechanisms that will provide long-term and reliable funding to conserve and sustainably manage the marine and coastal resources and the environment in each participating country and territory". The CBF manages approximately USD 70 million in funds, and the resources are mobilised through two financial instruments, an endowment fund and a sinking fund.

The CBF endowment fund, approximately USD 43 million, is the permanent funding source to the national conservation trust funds in the Caribbean region (including MEPA). It makes grants to the national trust funds, which administer the resources. To receive funds from the CBF, the national trust funds must have additional sustainable funding sources in place that are at least equal to the level of funding requested from the CBF. Launched in 2018 and with an expected duration of five years, the sinking fund is approximately USD 26.5 million, and focuses exclusively on funding ecosystem-based adaptation activities. A regional approach to CTFs has several advantages. It leverages economies of scale to reduce administration costs, can provide a convenient mechanism for donors looking to have regional impacts, and can catalyse the development of financing mechanisms and capacity at national levels. Regional CTFs can also facilitate co-ordinated action on shared environmental and sustainability issues, which is especially important for the ocean.

Another CTF is the Banc d'Arguin et de la Biodiversité Côtière et Marine (BACoMab) trust fund in Mauritania, which funds multiple activities related to the conservation and sustainable use of ocean resources. Founded in 2009, BACoMaB supports the management activities in the Banc d'Arguin and Diawling national parks as well as marine patrols, environmental education, development of governance systems, and restoration. The fund has a total capitalisation of EUR 26.6 million and in 2018, it spent EUR 460 000 on management activities (BACoMaB Trust Fund, 2019[31]). The ecosystem services provided by Banc d'Arguin National Park are essential to maintaining the fish stocks of the region. Consequently, under the sustainable fisheries partnership agreement between Mauritania and the European Union (EU), the BACoMaB has received approximately EUR 1.9 million for the delivery of these services (OECD, 2017[32]; BACoMaB Trust Fund, 2019[31]). Due to its flexibility, the trust fund has been able to act as both donor co-ordinator for the area and as an intermediary in an international PES scheme while distributing resources across a range of context-specific actions and financial mechanisms.

These two examples highlight the ability of trust funds to generate finance for the conservation and sustainable use of the ocean. The flexibility that CTFs have in leveraging different financial mechanisms allows them to co-ordinate both donors and stakeholders to fund a range of different actions to achieve a unified goal. For this reason, CTFs can play an important role in addressing complex problems in the ocean economy, such as marine plastics, that require a broad range of actions over several sectors. CTFs also can be used to co-ordinate the international community and leverage several different international mechanisms, such as debt for nature swaps, impact investments and more traditional grant funding (Chapter 4).

Payments for Ecosystem Services: Fostering coastal resilience via blue carbon payments

Coastal regions are increasingly affected by climate change. Sea-level rise, for example, is expected to cause USD 1.7-5.5 trillion in damage from coastal flooding over the 21st century (OECD, 2019[33]). Climate change is also expected to increase the severity and frequency of extreme weather events, which will severely impact coastal communities (IPCC, 2019[34]). Developing countries, with their relatively large coastal populations, are therefore particularly at risk from the impacts of climate change. Increasing the resilience of coastal areas, their populations and their industries to the impacts of climate change is a key policy priority to develop the ocean economy sustainably.

Governments play a central role in incentivising the development of coastal areas and, as such, should take the lead in ensuring that these developments are resilient to the impacts of climate change. The first step in this process is to understand how existing and planned coastal infrastructure and development are exposed to the impacts of climate change and environmental degradation. Otherwise, governments risk incentivising projects that will not be viable in the long term. Beyond new developments, ensuring coastal areas and communities are resilient to environmental risk is also essential.

Coastal ecosystems such as mangroves, salt marshes and seagrasses provide a range of highly valuable ecosystem services and can enhance coastal resilience by reducing the impacts of coastal flooding and erosion. Mangroves in particular provide significant benefits to adjacent communities, including reducing the impact of coastal flooding events by dissipating the energy from waves. For example, it is estimated that mangrove areas in the state of Florida in the United States avoided USD 1.5 billion in flooding damage due to Hurricane Irma (Narayan et al., 2019[35]). Relatedly, the loss of mangroves in Philippines has exposed 267 000 people to an increased risk of annual flooding. Restoring mangrove areas lost since 1950 in Philippines would provide an estimated USD 450 million in flood protection benefits annually (Menéndez et al., 2018[36]). These ecosystems can also sequester large amounts of carbon, thereby contributing to mitigating climate change and in some cases, potentially providing a source of revenue.

Mangroves occur predominately in tropical regions, which often have large and vulnerable coastal populations, and provide significant local-scale benefits such as fuel and food. Indonesia, for example, has both the largest coastal population in the world and also the largest extent of mangroves (~3 million hectares). Regions featuring mangroves generally are also characterised by high levels of poverty and consequently local populations can be heavily reliant on the ecosystems services they provide. In the Gazi Bay in Kenya, an estimated 80% of the local residents derive their livelihood directly from fishing-related activities, which are linked to mangroves due to essential role mangroves play in providing a nursey for maturing fish (Wylie, Sutton-Grier and Moore, 2016[37]). The ecosystem services provided by mangroves have been estimated at well over USD 100 000 per hectare, although location and other factors affect their estimated value (Himes-Cornell, Grose and Pendleton, 2018[38]).

The importance of mangrove ecosystems to the global carbon pool and the potential of mangrove restoration to mitigate greenhouse gas (GHG) emissions present a considerable opportunity to generate revenue through blue carbon projects. The loss of mangroves and other coastal ecosystems results in estimated emissions of 0.25-1.05 PgC a year, with mangroves containing approximately half of the global blue carbon pool (Pendleton et al., 2012[39]). Further, mangroves have high productivity, producing 10-15% of coastal sediment carbon storage despite occupying only 0.5% of the global coastal area (Alongi, 2014[40]). Globally, mangroves are being lost at a rapid rate – –around a third of mangroves have been lost to date – with serious consequences for both local communities and global climate change (Alongi et al., 2015[41]). Pendleton et al. (2012[39]) estimated the economic impacts of the emissions from mangrove loss to be USD 6-42 billion a year. The development of aquaculture has driven much of this loss. All four countries selected for Sustainable Ocean Economy Country Diagnostics aim to further expand the aquaculture sector, which potentially may lead to further loss of mangroves.

While opportunities for blue carbon projects vary across the four countries, Indonesia and Kenya feature extensive mangroves. The Indonesian mangroves are being lost at a rate of about 1% year leading to emission of around 26 400 gigagrammes (Gg) of carbon dioxide (CO_2) per year (Alongi et al., 2015[41]). Indonesia recognised the importance of mangrove conservation with the creation of the National Strategy for Mangrove Ecosystem Management in 2012, the Indonesian Blue Carbon Strategy Framework, and the low carbon development strategy that acknowledges the need to conserve mangroves. Several blue carbon projects are underway around the world. Box 3.4 describes one of these, the Mikoko Pamoja blue carbon project, in Gazi Bay, Kenya.

> Box 3.4. Mikoko Pamoja blue carbon project in Gazi Bay, Kenya
>
> - The Mikoko Pamoja blue carbon project in Gazi Bay, Kenya is a community-led project that uses mangrove restoration to generate revenue from the voluntary carbon markets. The project area includes 117 hectare (ha) of nationally owned mangroves. The mangroves provide several important ecosystems services to the local community such as food provision (by acting as a nursery for fish), recreation, tourism and fuel. However, the collection of timber, particularly for building, has led to degradation of the mangrove area.
> - Mikoko Pamoja has entered into an ecosystem services agreement with Plan Vivo, a certification body and foundation based in the United Kingdom. As part of the project, a non-native timber species was planted away from the mangroves as community forest to replace the supply of fuel wood and construction timber previously obtained from the mangrove area. Between 2013 and 2018, the project has been issued credits for 8 068 tCO_2 of emissions avoided, which has generated USD 61 534. This money has been used to maintain the project activities (including one full-time staff member) and fund two community development projects related to health and sanitation. In 2017, the project received the UNDP Equator Prize, which is awarded biennially "to recognize outstanding community efforts to reduce poverty through the conservation and sustainable use of biodiversity".
>
> Source: Huxham (2013[42]) *Mikoko Pamoja: Mangrove Conservation for Community Benefit*, Plan Vivo Foundation, Edinburgh, https://planvivo.org/docs/Mikoko-Pamoja-PDD_published.pdf and Mwamba et al. (2018[43]), *2017-2018 Plan Vivo Annual Report: Mikoko Pamoja*, Plan Vivo Foundation, Edinburgh, https://www.planvivo.org/docs/2017-2018_Mikoko-Pamoja-Annual-Report-Final-public_.pdf

Blue carbon projects aim to demonstrate emissions removals or avoidance through restoration activities to generate carbon credits; these credits can then be sold on either the compliance or the voluntary markets to generate revenue.[2] These include mechanisms such as the clean development mechanism (CDM) under the Kyoto protocol and the Reducing Emissions from Deforestation and Forest Degradation (REDD+) initiative under the United Framework Convention for Climate Change. There are several different voluntary market standards, among them the Verified Carbon Standard (VCS) and the Plan Vivo standard. Several blue carbon projects have already been certified either under VCS or Plan Vivo standards, but no projects have as yet been certified under the compliance standards (Wylie, Sutton-Grier and Moore, 2016[37]). This is likely related to the greater flexibility and lower transaction costs associated with the certification for the voluntary carbon markets.

The greater flexibility makes voluntary standards more attractive to emerging offsets approaches like blue carbon, and the lower transaction costs mean small-scale projects can be certified. Compliance markets have more rigorous standards of monitoring, reporting and verification that are challenging for smaller projects to meet in a cost-effective way. Under the CDM, for example, a project needs to sell at least 5 000 tCO_2 to justify the transaction costs, which puts this mechanism out of reach for many small-scale coastal projects (Kollmuss et al., 2008[44]). Among the countries for which Sustainable Ocean Economy Country Diagnostics were conducted, projects in Kenya and Indonesia have been certified successful under the

Plan Vivo and VCS standards, respectively. The impacts of these projects have been significant: the Yagasu project in Ache and North Sumatra provinces of Indonesia protects 25 000 ha of forest and restores a further 5 278 ha, resulting in annual emissions reductions of 120 706 tCO$_2$e (VCS, 2019[45]). Like the Mikoko Pamoja blue carbon project (Box 3.4), Yagasu has had significant positive social impacts by improving the livelihoods of over 9 000 people through employment, increased income and capacity building.

Given how closely linked the development of aquaculture is with mangroves, understanding how to leverage blue carbon projects to improve the sustainability of the aquaculture sector is important. The Markets and Mangroves (MAM) project in the province of Ca Maua on the Mekong delta in Viet Nam offers an interesting approach. The area has a large shrimp farming industry and the project encompasses 3 371 ha, of which 1 715 ha are mangroves (Wylie, Sutton-Grier and Moore, 2016[37]). MAM supported local shrimp farmers to gain organic certification and as part of that process, they must agree not to remove any more mangroves. Further, the project required enrolled farmers to maintain mangrove cover of 50% on their properties and facilitated the restoration of mangroves where required. Certified organic shrimp receive a price premium (of approximately 10%) on European and United States markets, and increased mangrove cover has been shown to improve the productivity of shrimp farms. Consequently, enrolled farmers saw significant profit increases. While carbon finance was intended to play an important role in the MAM project, the benefits of organic certification have so far proved sufficient to engage stakeholders. Developing carbon finance alongside certification, however, could further increase incentives to farmers to take part in this kind of programme.

Smaller-scale projects funded on voluntary carbon markets are a good opportunity for small island developing states (SIDS). Antigua and Barbuda, for example, has less than 1 000 ha of mangroves. Given the transaction costs associated with carbon compliance markets, it is both unattractive and potentially unprofitable to pursue carbon credits through these mechanisms. The experience of the Mikoko Pamoja in Kenya, however, suggests that voluntary markets can be used successfully for projects of smaller scale that could subsequently be enhanced through mechanisms such as organic certification, if aquaculture already exists in these areas.

To date, it has proved challenging to operationalise blue carbon projects at sufficient scales to generate significant revenue in countries with extensive mangroves, such as Indonesia. Under these circumstances, a more viable option could be to pursue certification and carbon finance through the compliance markets. Many developing countries, among them Indonesia, have well-developed programmes as part of the United Nations REDD initiative. Mangroves are a good candidate for inclusion in these programmes due to their rate of conversion and associated emissions. Understanding either how to adapt the systems developed for REDD+ to include blue carbon from mangroves or utilise existing capacity to create new systems may be a good way to significantly lower the establishment and transactions costs of these projects.

Technical challenges remain, including accurately measuring below-ground carbon sequestrations (where the majority of carbon is stored in mangrove ecosystems), accurately mapping mangrove extents and best practice for mangrove restoration (Macreadie et al., 2019[46]). In recognition of these challenges, several regional and international platforms have emerged. These include the Blue Forests initiative, co-funded by the Global Environment Facility; the Mangroves For the Future, a platform of the International Union for Conservation of Nature (IUCN) and UNDP; and the Blue Carbon Initiative of Conservation International, the International Oceanographic Institute and IUCN; all aim to foster research, information exchange and the development of best practice for mangrove restoration (and other ecosystems). It is essential to support and strengthen these efforts if developing nations are to successfully mobilise international carbon finance for the conservation and sustainable development of marine ecosystem services stemming from the ocean.

Finally, beyond carbon payments, there is an increasing role for innovative insurance mechanisms to protect and restore high-value coastal ecosystems. These include parametric insurance for coral reefs and beaches, in which a payout is triggered by an agreed parametric trigger (e.g. a storm reaches a certain windspeed) rather than by specific valuation of damages (Chapter 4). However, local-scale models of risk and benefit flows specific to individual ecosystems are essential for insurance instruments, and the data are lacking for many areas. This data paucity is a barrier to their broad use, and uptake of these instruments has been limited thus far. To increase the role of financial markets in financing coastal conservation and restoration, improved human and technological capacity for data collection and analysis is needed to build the kinds of risk models that the finance industry requires.

Policy incentives and finance mechanisms for sustainable fisheries and aquaculture

Sustainable fisheries

The fisheries sector is important to the economies of many developing countries and to the livelihoods of their populations (Chapter 2). The vast majority of the estimated 40.3 million fishers in the world work in small-scale and artisanal fisheries in developing countries (FAO, 2018[47]). The fisheries sector relies squarely on healthy, well-functioning marine ecosystems for its long-term sustainability. Yet unsustainable fishing has depleted fish stocks globally, resulting in an estimated USD 83 billion in forgone benefits annually (World Bank, 2017[48]).

Illegal, unreported and unregulated (IUU) fishing remains a serious problem, accounting for as much as 26 million tonnes of fish and costing the global economy up to USD 23.5 billion annually (Agnew et al., 2009[49]). IUU fishing has several important impacts: it can harm law-abiding fishers by creating unfair competition and cutting profitability weaken food security in countries that rely on local seafood; undermine fisheries management efforts by adding pressure that is difficult to quantify.

Developing countries often lack the legal frameworks and resources to monitor and police their water and hence are particularly vulnerable to IUU fishing (Hutniczak, Delpeuch and Leroy, 2019[50]). IUU fishing, therefore, not only degrades the integrity of marine ecosystems. It also puts strains the resources available to countries to combat it. In their recent study of the regulatory gaps in OECD and key partner countries, Hutniczak, Delpeuch and Leroy (2019[50]) find that countries with higher GDP per capita tend to have more developed legal frameworks for addressing IUU fishing . Therefore, an exchange of knowledge and experience among OECD and non-OECD countries could help developing nations to address IUU fishing.

Increasing the resources available to developing countries for fisheries monitoring and management is also key to making fisheries sustainable and growing the ocean economy. The experience of South Africa is an example. An integral part of its ocean economy strategy is to develop a National Ocean and Coastal Information System and extend its earth observation capacity (South Africa Department of Environmental Affairs, 2019[51]). In December 2018, South Africa launched its ZA-cube2 satellite, which revealed the presence of vessels and their exact locations in coastal waters. The data immediately contributed to enhance the South African Integrated Vessel Tracking System), allowing, in its first month in operation, the tracking of drug-carrying vessels in Port Elizabeth and the seizure of drugs worth ZAR 720 million (South Africa rand), equivalent to about USD 46 million (South Africa Department of Environmental Affairs, 2019[51]).

Developing the institutions and systems required to sustainably manage and monitor fisheries requires significant resources. Economic instruments and other fiscal policies can play a key role in enhancing the sustainability of the fisheries sector and in generating revenue. For instance, fees and charges are used extensively in the fisheries sector to raise revenue and control access to resources. In most cases, countries issue a licence to operate in particular fisheries in exchange for a fee. (The OECD PINE database includes information on fisheries-related fees in 43 countries.) The fee is usually based on the size of boat and type of gear being used, the species targeted, and, in some cases, whether the boat is foreign or domestically owned. These fees can generate quite substantial revenue. For instance, fees for fishing licences in Indonesia generated IDR 448 billion (approximately USD 31.7 million) in 2018 for example. For SIDS, which often have small populations and very large EEZs, fees relating to fisheries produce a substantial revenue stream and make up a significant proportion of total government revenues (Table 3.7). This is particularly true for Pacific Island countries and territories; Kiribati, for example, raised approximately USD 117 million in 2015 through fisheries access fees, an amount equivalent to USD 1 044 for every person in the country (Gillett, 2016[52]).

Providing access to foreign vessels in exchange for monetary compensation is another mechanism for raising revenue for the sustainable management of fisheries. Cabo Verde, for example, received access fees averaging EUR 839 361 a year from 2014-17 from vessels and from the EU Sustainable Fisheries Partnership agreement. However, only EUR 500 000 is ring-fenced for the fisheries sector, with the majority going directly to the central budget (European Commission, 2018[53]). Given the particular abundance of high-value marine species such as tunas in the Pacific Island countries and territories, it is unlikely other nations can replicate the level of fees generated. Further, if fees are contingent on catching migratory species, as is the case in Cabo Verde, revenue streams can be unreliable from year to year, making it difficult to plan budgets effectively.

Table 3.7. Revenue from access fees for foreign fishing vessels in Pacific Island countries and territories in USD, 2014

Country or /territory	Access fee revenue	Access fee revenue per capita	Access fee revenue per km^2 EEZ	Total access fee revenue as % of government revenue
Cook Islands	8 473 500	554	4.61	11.40%
Micronesia	47 518 000	462	5.96	20.90%
Fiji*	555 815	1	0.43	0.04%
Kiribati	116 040 984	1 044	32.69	75.00%
Marshall Islands	16 920 802	310	7.94	16.40%
Nauru	15 852 459	1 487	49.54	13.70%
Niue	635 815	424	1.63	3.30%
Palau	3 620 586	203	5.76	3.30%
Papua New Guinea	85 019 455	11	27.25	1.70%
Samoa**	555 814	3	4.63	0.30%
Solomon Islands	27 963 558	45	20.87	7.20%
Tokelau	9 050 000	7 762	31.21	52.50%
Tonga	627 858	6	0.9	0.40%
Tuvalu	14 777 814	1 321	16.42	58.30%
Vanuatu	1 759 112	6	2.59	1.00%

Note: *No foreign fishing allowed in its EEZ, but payments under the terms of the South Pacific Tuna treaty with the United States. ** No foreign fishing allowed in its EEZ. Source: Adapted from Gillett (2016[52]), *Fisheries in the Economies of Pacific Island Countries and Territories*, https://www.spc.int/sites/default/files/wordpresscontent/wp-content/uploads/2016/11/Gillett_16_Benefish-fisheries-in-economies-of-pacific-countries.pdf.

StatLink https://doi.org/10.1787/888934159411

Where pelagic fisheries are relatively underexploited, as in Kenya, and where domestic capacity to exploit the stocks is low, allowing paid access to foreign vessels can help raise funds that could then be used to further develop domestic fishing capacity. Many countries also directly support fisheries through support for fuel and the manufacture and distribution of fishing gear and fishing vessels. However, support that lowers the cost of inputs in general, and of fuel support in particular, is considered the most likely to lead to overfishing and IUU fishing and the least likely to deliver income benefits to fishers (OECD, 2017[54]).

While there is understandable pressure to improve global food security and provide livelihoods for fishing communities, governments and investors should focus on smart approaches that benefit fishers more (and in the longer term) and are not likely to encourage unsustainable fishing. It is essential to ensure that domestic management and monitoring are improved concurrently with improved capacity. Otherwise, any growth in the industry could be short-lived, as stocks could decline with increased fishing effort. Developing countries, therefore, must strike a delicate balance in their investments in the domestic fishing industry and in the tools and institutions to effectively manage their fisheries resources sustainably.

Growing the value of the fisheries sector

Seafood processing is also an important industry. Millions of people, particularly women, are involved in artisanal fish processing (OECD/FAO, 2019[55]). Assessing the seafood production value chain in developing countries is challenging as they often lack post-harvest facilities such as drying equipment as well as ice plants and cold storage facilities. Such installations can add value to the seafood product, increase the selling price and reduce the post-harvest losses that occur in these fisheries (Rosales et al., 2017[56]). When no storage facilities with ice are available in ports, for instance, fishers may have to either sell their catch too cheaply or lose it to spoilage. Developing more value added for the entire fish production will therefore highly dependent on enhancing the post-harvest processing chains.

Measures that add value, such as certifications, are seen to offer the greatest potential for increasing the socio-economic benefits from fishing, together with interventions to reduce post-harvest losses and to reduce by-catch (Carneiro et al., 2019[57]). Certification and eco-labelling can lead to price premiums for more sustainable fishing practices and to the prioritisation of value over quantity in seafood production (FAO, 2017[58]). Certification also makes an important contribution to greater traceability of fish products, ensuring that specific legal requirements are met along the value chain and reducing trading of illegally fished fish.

Sustainable aquaculture

Growing the aquaculture sector in coastal areas is an opportunity for developing countries to also grow the ocean economy and increase food security, as Chapter 2 notes. However, this needs to be managed effectively to ensure that growth in this sector is not at the expense of coastal ecosystems. For example, coastal production systems such as shrimp that rely on the creation of salt ponds can have negative impacts if they drive the conversion of high-value coastal ecosystems, such as mangroves. In addition, high-intensity, closed circulation systems require high levels of capital and require significant energy inputs, and thus may not be the best option for developing countries given their potential community-level impacts. Economic incentives to grow aquaculture should consider the environmental effects of expanded production and encourage production systems that best fit the country's environmental and socio-economic context.

Historically, the expansion of aquaculture has been associated with some negative environmental consequences such as land use change, pollution from excess feed and waste products, and the introduction of invasive species (Diana, 2009[59]). However, conclusions regarding its impacts vary widely. For example, some research has attributed only an estimated 10% of mangrove loss to aquaculture expansion globally (Diana, 2009[59]), while other evidence indicates that expansion of shrimp ponds has been the primary cause of mangrove loss in Thailand (Sampantamit et al., 2020[60]). At the same time, most aquaculture (69.5%) is of fed species, and the most nutritious sources of feed for aquaculture are fish oil and fishmeal. The use of fish oil and fishmeal in fish feed is declining in favour of alternate sources, predominately oil seed and other crops (FAO, 2018[47]). Understanding and accounting for the environmental impacts of feed production are essential to ensure aquaculture growth is sustainable.

Many developing countries are targeting growth in the aquaculture sector and using subsidies and other forms of support to encourage its development. In South Africa, some aquaculture businesses are eligible for government grants of up to ZAR 20 million (USD 1.1 million) under the aquaculture development and enhancement programme of the Department of Trade and Industry. The development of environmentally sustainable shrimp and prawn aquaculture is also a component of Thailand's *Twelfth National Economic and Social Development Plan (2017-2021)* (Office of Prime Minister of Thailand, 2017[61]). The aquaculture sector is eligible for a range of tax incentives and non-tax incentives under the country's Investment Promotion Act. The sector is also linked to a special economic zone (SEZ), a term often used for tax breaks for new investments in certain industries in a particular region. To ensure that this development is sustainable, countries must understand the environmental impacts of this increased production, which calls for a strong environmental impact assessment process. A comprehensive spatial planning process also can help to avoid ecologically sensitive areas and minimise additional impacts (Henriksson et al., 2019[62]).

Data on the impacts of potential aquaculture development are essential for selecting the aquaculture production system best suited to local environmental and socio-economic conditions. Low input systems, such as the production of molluscs and seaweed, for example, require smaller capital inputs and generally produce lower value products (in the case of seaweed). But such systems can be implemented at community or individual level. Whereas more intensive systems, produce more, but require higher capital and energy inputs, reducing their suitability for many developing country contexts. At the extreme end, large recirculating saltwater aquaculture systems can produce fish using 1-3% of the water used in traditional aquaculture, but these systems require sophisticated technology and high capital inputs with payback periods of about eight years (Bregnballe, 2015[63]). Choosing the most appropriate system for a given context is therefore essential to ensure future aquaculture production is socially and economically sustainable.

Policy incentives and finance mechanisms for sustainable tourism

Another important sector to promote the conservation and sustainable use of the ocean is tourism, a key pillar of the economy in many developing countries (Chapter 2). The rapid growth of the tourism sector over the last 20 years is predicted to continue, with sustainability issues becoming increasingly important (OECD, 2020[64]). Relying as they do on the recreational value of beaches and clean waters, marine and coastal tourism are highly dependent on the quality of natural ecosystems to attract visitors. Yet unmanaged tourism is contributing to ecosystem depletion and fragility, jeopardising the sector's own economic sustainability. Key challenges include the beach degradation resulting from sand harvesting, mangroves deforestation and an ever-growing coastal population that is putting pressure on coastal ecosystems. Climate change vulnerability is also a risk for countries that rely on tourism, particularly so for small island states where tourism is also the largest sector of the national economy and the largest employer (Scott, Hall and Gössling, 2019[65]). In Antigua and Barbuda, for example, tourism represents nearly 60% of the GDP. The tension between the promotion of commercial activities to foster economic growth and the need to address environmental concerns is a key issue for most developing countries.

Globally, tourism also has significant adverse environmental impacts: it directly contributes 5% of global GHG emissions and increasing the pressures on domestic freshwater supplies, food systems and waste disposal systems (OECD, 2018[66]). The economic and environmental ramifications of overuse by tourism of high-value marine ecosystems can be significant. An example is Maya Bay in Thailand, which closed in 2018 after a rapid increase in visitor numbers caused the severe degradation of both marine and terrestrial ecosystems. The bay will remain closed until at least June 2021. Boracay island, a popular tourist destination in Philippines, was closed for six months in 2018 after it emerged that the vast majority of businesses and residents lacked wastewater permits and in many cases were discharging sewage directly into the sea (Reyes et al., 2018[67]). The island closure affected the employment of 35 000 people and cost an estimated 20 billion pesos (approximately USD 395 million). Reducing the negative impact of tourism on coastal zones, therefore, is a priority area for the sustainable ocean economy and one that will bring long-term economic and environmental benefits.

The OECD (2018[66]) identified the shift to sustainable tourism as a tourism sector megatrend to 2040. Megatrends are large-scale social, economic, environmental and technological changes that are likely to have profound long-term consequences for a sector. Sustainable tourism is an opportunity for developing countries to grow a key sector of the ocean economy while maintaining and enhancing natural capital to deliver multiple benefits to society and nature. This section highlights policy instruments and mechanisms available to developing countries and examples of how they are being applied and can be used to generate revenue and finance the transition to sustainable tourism.

Tourism-related taxes for sustainable development

Tourism-related taxes can create incentives for sustainable tourism. Revenue generated by these policy instruments, if earmarked, can also be used to finance activities related to the conservation and sustainable use of the ocean and to support local and vulnerable communities dependent on the ocean. Taxes can also help to counter financial leakages in the tourism industry[3] that reduce the economic benefits derived from tourism in developing countries; and have reduced the effectiveness of tourism on poverty alleviation in developing countries (World Tourism Organization, 2002[68]). Tourism-related taxes may include a specific mechanism to direct revenue towards relevant activities through as earmarking or credits for investments in marine conservation and sustainable use activities. Revenue from well-designed taxes on the tourism industry and the goods and services consumed also can help developing countries manage the impacts of development, with most of the tax burden falling on generally wealthy foreign nationals. However, such taxes also can reduce the competitiveness and attractiveness of destinations to both investors and tourists (OECD, 2014[69]).

A good practice example is the corporate social responsibility tax introduced in 2014 in the Seychelles. It was introduced in 2014. The tax, a 0.5% levy on the business turnover, includes a credit of up to 0.25% of turnover for investments in community projects, including biodiversity conservation, and has been taken up by several local hotels (UNDP, 2019[70]).

As highlighted in Chapter 2, the cruise industry is an important sector for small island developing states, and raising levies on passengers is a useful mean to generate revenues to address sustainability challenges. Antigua and Barbuda imposes a USD 1.50-per person levy on cruise ship passengers that generated approximately USD 1.2 million in 2018. This revenue is transferred directly to the national solid waste management agency to recover the costs of dealing with the waste generated by cruise ships and their passengers. Raising the levy to include a component related to the management of marine biodiversity could be a good way to generate additional resources. The equivalent fee levied by the Seychelles is just over USD 7; the UNDP Biodiversity Finance Initiative has recommended it be raised to USD 20. This recommended fee is also significantly higher than the current Antigua and Barbuda levy, suggesting there are untapped opportunities to raise more revenue from this tourism-related taxes.

Levying fees and taxes on cruise ships (and their passengers) has been challenging in the Caribbean, as the islands, individually, are in a weak negotiating position with the cruise companies and are in fierce competition with other islands for passengers. As a result, there has been a race to the bottom, with taxes on cruise passengers in the Caribbean ranging from USD 1.50 to USD 18. Greater co-ordination among the Caribbean nations could strengthen their collective negotiating position and lead to increased government revenues to support environmentally sustainable growth in cruise tourism (MacClellan, 2019[71]). However, given the extent of environmental impacts from cruising, both in and out of port, taxes alone may not generate sufficient revenue to ensure sustainability. Other regulatory measures are likely required.

An example of tourism-related taxes as an instrument to support environmental protection is the Iceland accommodation tax, highlighted in *Tourism Trends and Policies 2014*, an OECD review of taxation of the tourism industry in 30 OECD countries and partner economies (OECD, 2014[69]). The tax partially finances the Icelandic Tourist Site Protection Fund, which was established in 2011 and allocates 40% of its funds to the Environment Agency of Iceland for national parks. Since its inception, the Fund has supported more than 750 projects and in 2017, allocated ISK 700 million (Icelandic króna). (OECD, 2018[66]). Before the establishment of the fund only ISK 50 million was available for the parks.

Tourism-related subsidies

Government subsidies can also be used to promote more sustainable tourism. Kenya, for example, offers a customs duty exemption for equipment relating to the creation of wastewater treatment plants for hotels.[4] Thus, hotels in Kenya are incentivised to invest in wastewater treatment plants, which potentially reduces both the environmental impact of the hotels and the pressure new hotel developments place on local infrastructure.

In many cases, government support to the ocean economy takes the form of an SEZ. Their use has grown rapidly in the last six years: in 2019, 4 772 SEZs were operating in developing economies (UNCTAD, 2019[72]). SEZs can be used to promote investment in large-scale infrastructure projects for the ocean economy such as ports or fish processing facilities. SEZs have been criticised for negative environmental impacts that stem from relaxed regulations and from countries' prioritisation of development in the short term over the long-term sustainability of the projects (UNCTAD, 2019[72]).

For developing countries, better integration of environmental sustainability objectives into the conditions for receiving subsidies in SEZs, combined with a strong process of assessing the impacts of development, is key if they are to make the most of subsidies to grow the sustainable ocean economy. Indonesia provides an example. It uses SEZs and a range of subsidies to encourage new investments in less-developed areas. Four tourism SEZs with a legal basis (i.e. specific regulation and institutional structures) were operating in Indonesia 2019: Mandalika (Lombok), Mortoai (Maluku), Tanjung Kelayan (Belitung) and Tanjung Lesung (Banten). The subsidies take the form of tax holidays, excise and import duty exemptions. and other tax allowances designed to reduce the cost of the project. However, the pace and extent of the tourism expansion anticipated in these areas raise some concerns regarding the sustainability of the growth (Ollivaud and Haxton, 2019[73]). For example, in Bali, tourists use significantly more water (up to 200 L per day) than do residents (up to 60 L per day for rural and 120 L per day for urban residents), putting a strain on local infrastructure (Ollivaud and Haxton, 2019[73]). Further, Indonesia is the world's second largest contributor to ocean plastic waste, which increasing numbers of tourists could exacerbate (Jambeck et al., 2015[2]).

Private sector tourism-related concessions and community co-management of MPAs

Ensuring a sustainable source of financing for MPAs is essential for the conservation and sustainable use of marine resources in the ocean economy. In some – but not all – cases, the cost of management is covered by a combination of government allocations and user fees (OECD, 2017[4]) (see above). Partnering with the private sector to offer concessions for exclusive access to resources can provide MPAs a sustainable source of financing in the form of fees paid by the private sector. Providing exclusive access rights creates a sense of ownership in the individual companies that can inspire good behaviour and environmental stewardship. However, private concessions also can cause conflict with local communities if they lead to the perceived or actual loss of rights to certain resources. Comprehensive stakeholder consultation and measures to address distributional impacts of the concession are key to the success of concessions, particularly in areas with vulnerable and indigenous communities.

The tourism sector is potentially well suited to MPA concessions since the desirability of a destination is often directly linked to environmental quality. Relatedly, issues relating to excess tourism, such as overcrowding and environmental degradation, can negatively impact the competitiveness of a destination. The tourism and travel sector is expanding and international arrival grew by 3.8% in 2019 (World Tourism Organization, 2019[74]). Wildlife tourism alone, defined as viewing or experiencing animals in their natural habitat, directly contributed USD 120.1 billion to global GDP in 2018, representing 4.4% of tourism and travel sector GDP. Its share of the market is much higher in some areas than in others, for example representing 36.3% of the tourism and travel sector in Africa (World Travel & Tourism Council, 2019[18]). A key challenge is to convert this growth in the tourism sector overall and the value of wildlife tourism into sustainable financing for marine ecosystems.

The size of conservation concessions can vary considerably, ranging from individual tour or dive operators to large resorts. Concession fees must be structured to ensure that the concession is an attractive investment and that the revenue generated is sufficient to cover the costs of the management. This can be a problem if the management authority has no experience of awarding concessions and may not be aware of their value or if the authority awards contracts for less than their market value because of pressure to act quickly. Unscrupulous private sector actors may also underreport revenues from their activities, reducing the resources generated by the concession. Concession fees based on visitor numbers can avoid this latter problem, but it may not be an effective approach in a marine area with no clear entry point (Iyer et al., 2018[27]). Beyond the fees they generate, concessions can also aid the conservation and sustainable use of marine resources through improving the management of an area (e.g. through increased resources such as staff), the development of infrastructure for visitors, development of the local communities, and the restriction of access (Iyer et al., 2018[27]).

In many regions, the local community has access to and utilises areas in an MPA. Granting exclusive access to a private entity and restricting local community access can create resentment. A transparent and open concession process that is based on clear legal frameworks is, therefore, key to ensuring that local communities are not unfairly excluded. Further engaging the local community and all other relevant stakeholders in the process of creating the concession is also important to ensure transparency and equity. The essential components of a successful concessions system, identified in UNDP guidelines for enacting concessions in protected areas, are illustrated in Figure 3.3 (Thompson et al., 2014[23]). Each of these components is present to some extent in all concession programmes, although this will vary considerably based on the size and complexity of the concession itself. Concession projects should be adaptable and improved as needed over time. Given how complex many projects can be, it is essential to incorporate and use the latest available technology and techniques to manage human, financial and natural capital efficiently. Adaptability will help concessions to generate revenue while maximising the environmental and social outcomes.

Figure 3.3. Components of a successful concession system

Components arranged around a central circle containing: Organisational support, Transparency, Clear & fair decisions, Continuous improvement. Surrounding components: Law & Policy, Data, Online info, EIA monitoring compliance, Staff, Standard contracts, Planning, Process & procedure, Fees cost recovery incentives.

Note: This figure highlights the main components of a concession system. All of these components will need to be present to some degree for the concession system to be a success.
Source: Adapted from Thompson et. al. (2014[23]), *Tourism Concessions in Protected Areas: Guidelines for Managers*, https://www.undp.org/content/undp/en/home/librarypage/environment-energy/ecosystems_and_biodiversity/tourism-concessions-in-protected-natural-areas.html.

Engaging the local community in the process of developing the concession can create a sense of shared ownership between the concession-holder and other stakeholders. Moreover, strong community support and engagement are essential for effecting management of protected areas (Watson et al., 2014[20]). A good illustration is the Misool Eco Resort (MER) in Indonesia, where private sector concessions and the local community have worked well together. In 2005, the MER signed a 25-year access agreement with the local communities in Batbitim, West Papua. It was signed by the heads of two families who hold customary ownership rights to the area and local officials from various government and community institutions. The tourism office issued a business permit and the municipal governor (*upati*) gave verbal support. The agreement grants exclusive access to the tourism operator to develop infrastructure and run dive operations and prohibited the removal of marine resources from a 40 000-ha area; it also includes cash payments, education and employment for the local community. A specific exemption from the prohibition on removing marine resources, valid for a two-week period once every two years, is included in the agreement to ensure the survival of the traditional fishing practices of the local people (The Nature Conservancy, 2010[75]). Since the founding of the MER, there has been a 250% increase in fish biomass in the area protected by community and MER-funded patrols, indicating concessional approaches can be effective for protected marine resources and creating sustainable growth in the tourism sector.

The Galapagos National Park (GNP) in Ecuador is another good example of how concessions can work to control access and generate revenue for the conservation and sustainable use of biodiversity. As noted in the subsection on fees and charges in MPAs in this chapter, the number of visitors to the GNP has grown from about 17 000 in 1980 to 241 800 in 2017, an increase of 1 422% (La Dirección del Parque Nacional Galápagos [Galapagos National Park Directorate], 2018[76]). Tourism accounted for approximately half the Galapagos GDP in 2013, and the increasing visitor numbers generate significant value for the economy of the island. But they could place increasing strain on habitat in the national park. To control the visitors, the GNP restricts the number of concessions licenses it grants based on assessment of the carrying capacity, environmental impacts and the GNP management plan. The Ministry of Tourism then must approve the plan for the number of licenses and once it does, the GNP accepts applications for concessions and after considering them, publicly announces the winning bids. All concessions expire annually and are not renewed if more than one breach of the concession conditions has been reported. In 2011, the GNP generated USD 15.5 million, of which concessions represented the third largest contributions of around 7%.[5] As the total revenues of the GNP exceed its operating budget of USD 14.4 million, concessions form a valuable part of the sustainable financing model for the protected area (Thompson et al., 2014[23]).

Policy instruments to curb marine pollution, including plastic

Managing the interface between the land and ocean is crucial for the sustainable ocean economy. Pollution from land-based production and consumption has significant adverse impacts on the ocean, for example. Insert text here on marine pollution (80% from land-based sources). Marine pollution includes chemical, light, noise and plastic pollution. Examples of chemical pollution are pesticides, fertilisers, oil, industrial chemicals and sewage. Light and noise also have adverse impacts on the ocean and marine species. Noise from ships, sonar devices and oil rigs, for example, cause disruption to marine mammals by affecting their communication, migration, hunting and reproduction patterns. Overall, it is estimated that approximately 80% of marine pollution originates from land-based sources.

As highlighted in Chapter 1, about 8 million tonnes of plastic enter the ocean every year; are estimated to cost society from USD 500 billion to USD 2 500 billion a year in lost ecosystem services; and result in losses of USD 13 billion a year to tourism and fisheries (OECD, 2018[77]; Beaumont et al., 2019[78]). Addressing marine plastics, therefore, is a policy priority for both developed and developing countries.

Marine plastic pollution affects both the marine environment and human health. For example, plastics discarded at sea (often fishing gear) can cause the deaths of large marine animals that become entangled, and gear can continue to ghost fish for several years, thus reducing fisheries production. Single-use plastics, such as drinking straws and carrier bags, also are consumed by marine animals, leading to their deaths in extreme cases. Further, global plastic production is expected to triple by 2050, with the increased consumption projected to be in countries that currently lack capacity to effectively manage their waste (World Economic Forum, 2016[79]). As microplastics accumulate in the marine food chain, the consumption of plastics by people, through seafood, is also becoming a public health concern. However, empirical evidence for health risks of microplastics consumption is currently limited (Koelmans et al., 2017[80]).

Jambeck et al. (2015[2]) identify coastal population and efficacy of solid waste management systems as key determinants of marine plastic pollution. The authors also cite some studies that point to China as the biggest contributor to marine plastic pollution globally followed by Indonesia, Philippines, Viet Nam and Sri Lanka. However, these estimates do not account for illegal dumping or the import and export of waste internationally, which are relevant for establishing a better understanding of the problem of plastic pollution (Miedzinski, Mazzucato and Ekins, 2019[81]).

The OECD (2018[77]) Environment Policy Paper on improving plastics management discusses the environmental implications of marine plastic waste and sets out several options to address it:

- Changes in product design: These can include the use of alternative materials in place of plastics or changes that reduce the need for plastics in the first place. Shifting to biodegradable or bio-based plastics could also reduce the environmental impacts of marine plastic pollution. However, such a shift could increase environmental burdens elsewhere, and biodegradable plastics may still persist in the environment as microplastics or nanoplastics if they do not fully break down.
- Improve waste management: Improving the collection and recycling of plastics can reduce the rate of at which plastic enters the ocean.

- Plastic clean up: Manually removing plastic waste from problem areas such as beaches and directly from the ocean itself will ameliorate the impacts of plastics in certain areas. However, manually removing plastic waste is likely to be expensive.

Interventions to address plastic waste should occur as early in the waste management cycle as possible. In many cases, reducing the creation of plastic waste to begin with is the most desirable option. Countries have used economic instruments (taxes and fees) to reduce demand for single-use plastics. There has been a sharp rise since 2015 in the number of policies to address single-use plastic, in particular policies to reduce the use of plastic bags (UN Environment, 2018[82]). In most cases, these policies include partial or total bans on the use of certain types of bags or impose levies on the consumption of plastic bags. Levies on plastics can be imposed on suppliers, retailers or consumers, and often take the form of a flat rate charge on plastic bags. Several developing countries, including Antigua and Barbuda, Cabo Verde, and Kenya, have put in place complete bans on the use of certain types of single-use plastic bags. However it difficult to assess the environmental effectiveness of these bans due to a lack of monitoring (UN Environment, 2018[82]).

Indonesia has imposed a levy of USD 0.015 on plastic bags in 23 cities. While it resulted in a large drop in plastic bag usage initially, the levy has but has encountered significant retailer and consumer resistance that has delayed implementation of the policy nationwide (Langenheim, 2017[83]). Levies and bans have had significant success in reducing usage of plastic bags in some other countries such as Belgium, Denmark and the United Kingdom, although no data exist on the impact of approximately 50% of the policies implemented globally (UN Environment, 2018[82]). Where bans and levies have been successful, they are often associated with strong awareness and education campaigns that target increased awareness of the impacts of plastic pollution and with alternatives to single-use plastics. This underscores the importance of public education around plastics and how they can be avoided in policy mixes designed to reduce plastic waste. In addition to awareness, ensuring affordable alternatives to single-use plastic and enforcement of bans or levies are essential conditions for success. Extended producer responsibility, whereby producers are given responsibility for the disposal of post-consumer products, could also incentivise the uptake of single-use plastic alternatives among manufacturers. Finally, a ban on single-use plastic can create opportunities for new business that can provide sustainable alternatives. Hence, reducing single-use plastic can be good for both the environment and the economy if policy instruments are managed carefully.

Addressing other types of marine pollution is equally critical. Various policy instruments are available, depending on the source of pollution. Taxes on the use of pesticides and fertilisers can help to prevent their excessive use (Sud, 2020[84]). Such a policy instrument relies on the polluter pays principle and can generate finance for marine pollution prevention and control, for instance to help fund infrastructure investments in the treatment, recycling and reuse of wastewater. Other ways to incentivise positive behaviour include an initiative by the South African Port of Durban that offers a 10% reduction on port fees to ships that use specific types of fuels and have therefore received of a Green Award. The Panama Canal Authority, in another example, has changed berth allocation policies to incorporate environmental considerations. Through its Environmental Premium Ranking, the Canal Authority recognises shipping companies that meet environmental criteria entitling them to obtain certain priority rights in canal transit. Commonly used criteria include a ship's nitrogen oxide emissions, the type of fuel used and other indices of energy efficiency (ITF, 2018[85]). More general approaches are also important, such as raising public awareness by educating people about preventing marine pollution through radio, television and social media and in classrooms.

Science, technology and innovation for sustainable ocean economies

Scientific and technological advances are expected to play a crucial role in improving both an integrated and evidence-based management of the ocean and the sustainability of ocean-based economic activities themselves (OECD, 2019[86]). For developing countries, such advances can contribute greatly to building national ocean strategies by enhancing understanding of national marine resources and enabling more effective monitoring of the activities taking place in national waters and beyond.

Understanding national marine resources

Each country needs information about its national, ocean-based resources to develop the ocean economy and enhance sustainability. Ocean scientific programmes often are the starting point for the development of national marine spatial planning (MSP). Sustainable and synergic uses of marine and coastal areas and resources benefit from the adoption of an ecosystem-based approach to MSP and integrated coastal zone management practices (UN Environment, 2019[87]).

Expanding scientific knowledge on the ocean opens up economic development opportunities. Recent advancements in ocean observation capacity and vessel tracking capabilities, for example, led to the development of new systems for mapping and monitoring coastal and marine environments. Such systems have proven to be useful, particularly when applied in sustainable fisheries management (e.g. the South Africa National Ocean and Coastal Information System). While these systems are becoming more affordable, in many cases, links with policy-makers, regulatory bodies, and police and naval forces need to be reinforced to ensure the enforcement of compliance.

As developing countries further develop their ocean strategies, it is worth highlighting two areas that can assist in making the most of ocean science, technologies and innovation: ocean economy knowledge and innovation networks and international co-operation initiatives.

Participating in ocean economy knowledge and innovation networks

Many countries are engaging in emerging ocean economy knowledge and innovation networks (OECD, 2019[86]). These initiatives, some building on decades of co-operation, strive to bring together a diversity of players – public research institutes, small and medium-sized enterprises, large enterprises, universities and other public agencies – to work on scientific and technological innovations in many different sectors of the ocean economy such as marine robots and autonomous vehicles, renewable energy systems, and biotechnologies. The actors then apply their innovations, often smart combinations of existing and/or new technologies, to tackle complex problems in ocean applications that range from ocean monitoring and marine ecosystems regeneration to aquaculture and marine renewable energies.

This is an opportune time for developing countries to benefit from projects led by these knowledge and innovation networks mainly based in OECD countries. The UN has declared 2021-30 the Decade of Ocean Science for Sustainable Development. Developing a sustainable ocean economy is the focus as well of Sustainable Development Goal 14 (conserve and sustainably use the oceans, seas and marine resources for sustainable development). Target 14.A encapsulates one of the main rationales for stronger co-operation between the ocean science and research and development communities on one hand and the developing countries on the other:

> "Target 14.A Increase scientific knowledge, develop research capacity and transfer marine technology, taking into account the Intergovernmental Oceanographic Commission Criteria and Guidelines on the Transfer of Marine Technology, in order to improve ocean health and to enhance the contribution of marine biodiversity to the development of developing countries, in particular small island developing states and least developed countries" (UN, 2019[88]).

In this context, several programmes are encouraging knowledge transfer and capacity building. Some examples are shown in Box 3.5.

> **Box 3.5. Networks that facilitate the sharing of scientific know-how and transfer of marine technology**
>
> The Commonwealth Blue Charter is an agreement by all 53 Commonwealth countries "to actively co-operate to solve ocean-related problems and meet commitments for sustainable ocean development" (Commonwealth Secretariat, 2020[89]). With support from the Commonwealth Secretariat, many countries are currently assessing the potential economic value of their oceanic waters, with small states often focusing on key sectors such as artisanal fishing, maritime transport and tourism (Roberts and Ali, 2018[90]). Ocean exploration and marine mapping projects, conducted in partnerships with United Kingdom oceanographic institutions, provide important contributions to the assessments.
>
> Building on a decade of different research projects in the Macaronesian region and working with Portuguese stakeholders and developing country partners, the Oceanic Platform of the Canary Islands established a real-time observational network for water and air quality monitoring. Monitoring is conducted in eight different ports (including Madeira, Azores, Canary Islands and Cabo Verde) to promote marine research and sustainable development. As part of this "ECOMARPORT project", technology transfer and specialised training are provided from the Canary Islands and through Cabo Verde to the rest of the Macaronesian archipelagos and Western Africa (PLOCAN, 2020[91]).
>
> The Atlantic International Research Centre (AIR Centre), supported by the Portuguese Science and Technology Foundation, is another distributed network for international co-operation. It fosters scientific and technology projects focused on ecosystems and sustainable ocean activities locally and globally and between developed and developing countries on both sides of the Atlantic. Major research themes bring together international teams, for example on the comparative analysis of bays and estuaries. Branch offices of the AIR Centre and affiliated institutions are based in Angola, Brazil, Cabo Verde, Nigeria, Portugal, South Africa, Spain and the United States. Additional collaborative and joint ventures in development have been initiated in Ghana, Namibia, Norway, Senegal and the United Kingdom (AIR Centre, 2020[92]).
>
> The Indian Ocean Rim Association (IORA) is the lead body for promoting regional collaboration across the Indian Ocean. The IORA Blue Economy Core Group aims to encourage South-South collaboration on the ocean economy of the IORA member states (Bohler-Muller, 2014[93]). The IORA provides a framework for academic co-operation and researcher exchanges in the area of blue growth, focusing on analysis of risks and threats to peaceful trade in the region such as territorial maritime disputes; maritime piracy; terrorism against ships, ports and other critical infrastructure; organised sea-borne crime and trafficking; and the potential impacts of natural or man-made disasters and extreme events. Other areas of focus include aquaculture and food security and may be expanded to include shipping and oil and gas exploration.

Joining forces regionally for joint ocean exploration and conservation and sustainable use of marine ecosystems

The development of more sustainable ocean economies can both contribute to and benefit from enhanced regional collaboration. As demonstrated by ongoing MSP efforts, cross-border co-operation can be crucial to make the most of scarce financial resources for cost-effective ocean mapping, joint marine exploitation and conservation efforts.

The co-operation between the Seychelles and Mauritius in the scientific mapping and delimitation of their continental shelf is a good case in point. In 2008, the countries made a joint submission to the Commission on the Limits of the Continental Shelf concerning the Mascarene Plateau, an extended continental shelf of approximately 396 000 km².kilometres. In 2012, they signed a treaty whereby they agree to jointly exercise their sovereign rights to manage and exploit the resources in the "joint zone" of the Plateau. Along with exploration, the agreement concerns environmental protection, marine resources management including fisheries and hydrocarbons, and, importantly, the equal sharing of all resources in the zone. Through a jointly managed authority that oversees the activities in the area, the Seychelles and Mauritius plan to issue licenses for oil exploration and exploitation (UNECA, 2016[94]).

Regional bodies also are actively working on ocean strategies, including in the Caribbean region where countries are very diverse and have different levels of institutional and financial capacity to harness opportunities linked to the ocean economy (UN, 2018[95]). Most of the ocean economy in the region is based on tourism, shipping, mining and, for a limited number of countries, artisanal fisheries. In 2019, the Organisation of Eastern Caribbean States (OECS) launched a Green-Blue Economy Strategy and Action Plan, the region's first, to combine green and blue economy strategies for sustainable development (Organisation of Eastern Caribbean States, 2019[96]). The OECS comprises 11 countries and territories in the Eastern Caribbean: Anguilla, Antigua and Barbuda, British Virgin Islands, Dominica, Grenada, Guadeloupe, Martinique, Montserrat, Saint Kitts and Nevis, Saint Lucia, and Saint Vincent and the Grenadines.

In another illustration of co-operation, ten governments signed the 2014 Hamilton Declaration on Collaboration for the Conservation of the Sargasso Sea to tackle the Sargassum challenges. The ten are Azores, Bahamas, Bermuda, British Virgin Islands, Canada, Cayman Islands, Dominican Republic, Monaco, United Kingdom and the United States. The Hamilton Declaration is the first instrument to establish an international body to conserve an ecosystem that lies primarily in an area beyond national jurisdiction (Sargasso Sea Commission, 2019[97]).

The ocean economy is also drawing interest at continental level. In 2016, the UN Economic Commission for Africa published a policy handbook to provide initial advice to African member states on mainstreaming the blue economy into their national development plans, strategies, policies and laws (UNECA, 2016[94]). The African Union also launched several initiatives, including its 2050 Integrated Maritime Strategy to provide a broad framework for the protection and sustainable exploitation of Africa's marine resources. A centrepiece of the strategy is creation of a Combined Exclusive Maritime Zone of Africa, a common maritime space intended to boost trade, protect the environment and fisheries, enable the sharing of information, and strengthen border protection and defence activities.

References

Agnew, D. et al. (2009), "Estimating the Worldwide Extent of Illegal Fishing", *PLoS ONE*, Vol. 4/2, p. e4570, http://dx.doi.org/10.1371/journal.pone.0004570. [49]

AIR Centre (2020), *AIR Centre - About Us (webpage)*, https://www.aircentre.org/ (accessed on 10 April 2020). [92]

Alongi, D. (2014), "Carbon cycling and storage in mangrove forests", *Annual Review of Marine Science*, Vol. 6/1, pp. 195-219, http://dx.doi.org/10.1146/annurev-marine-010213-135020. [40]

Alongi, D. et al. (2015), "Indonesia's blue carbon: A globally significant and vulnerable sink for seagrass and mangrove carbon", *Wetlands Ecology and Management*, Vol. 24/1, pp. 3-13, http://dx.doi.org/10.1007/s11273-015-9446-y. [41]

BACoMaB Trust Fund (2019), *Rapport Annuel 2018 [Annual Report 2018]*, https://bacomab.org/wp-content/uploads/2019/04/Rapport-annuel-BACoMAB-2018-2.pdf. [31]

Beaumont, N. et al. (2019), "Global ecological, social and economic impacts of marine plastic", *Marine Pollution Bulletin*, Vol. 142, pp. 189-195, http://dx.doi.org/10.1016/j.marpolbul.2019.03.022. [78]

Bohler-Muller, N. (2014), "The Big Blue: Regional networks across the Indian Ocean", *Human Sciences Research Council*, http://www.hsrc.ac.za/en/review/hsrc-review-november-2014/the-big-blue-networks-across-indian-ocean. [93]

Bohorquez, J., A. Dvarskas and E. Pikitch (2019), "Filling the data gap: A pressing need for advancing MPA sustainable finance", *Frontiers in Marine Science*, Vol. 6, http://dx.doi.org/10.3389/fmars.2019.00045. [21]

Bregnballe, J. (2015), *A Guide to Recirculation Aqauaculture*, Food and Agriculture Organization/EUROFISH International Organisation, http://www.fao.org/3/a-i4626e.pdf. [63]

Cabral, R. et al. (2018), "Rapid and lasting gains from solving illegal fishing", *Nature Ecology & Evolution*, Vol. 2/4, pp. 650-658, http://dx.doi.org/10.1038/s41559-018-0499-1. [99]

Caribbean Biodiversity Fund (n.d.), *Antigua and Barbuda: Marine Ecosystem Protected Area Trust*, https://www.caribbeanbiodiversityfund.org/antigua-and-barbuda (accessed on 2020). [28]

Commonwealth Secretariat (2020), *The Commonwealth Blue Charter (webpage)*, https://bluecharter.thecommonwealth.org/. [89]

Costanza, R. et al. (2014), "Changes in the global value of ecosystem services", *Global Environmental Change*, Vol. 26, pp. 152-158, http://dx.doi.org/10.1016/j.gloenvcha.2014.04.002. [11]

Diana, J. (2009), "Aquaculture production and biodiversity conservation", *BioScience*, Vol. 59/1, pp. 27-38, http://dx.doi.org/10.1525/bio.2009.59.1.7. [59]

Ehler, C. and F. Douvere (2009), "Marine spatial planning: A step-by-step approach toward ecosystem-based management", *IOC Manuals and Guides*, No. 53, UNESCO, Paris, http://dx.doi.org/10.25607/OBP-43. [5]

European Commission (2018), *Ex-post and Ex-ante Evaluations Study of the Sustainable Fisheries Partnership Agreement between the European Union and the Republic of Cabo Verde*. [53]

Expert Group for Aid Studies, S. (ed.) (2019), *Fishing Aid: Mapping and Synthesising Evidence in Support of SDG 14 Fisheries Targets*, https://eba.se/en/rapporter/fishing-aid-mapping-and-synthesising-evidence-in-support-of-sdg-14-fisheries-targets/11447/. [57]

FAO (2018), *The State of World Fisheries and Aquaculture 2018*, Food and Agriculture Organization (FAO), Rome, http://www.fao.org/3/I9540EN/i9540en.pdf. [47]

FAO (2017), "Seafood traceability for fisheries compliance: Country-level support for catch documentation schemes", *FAO Fisheries and Aquaculture Technical Paper*, No. 619, Food and Agriculture Organization, Rome, http://www.fao.org/3/a-i8183e.pdf. [58]

Gill, D. et al. (2017), "Capacity shortfalls hinder the performance of marine protected areas globally", *Nature*, Vol. 543/7647, pp. 665-669, http://dx.doi.org/10.1038/nature21708. [12]

Gillett, R. (2016), *Fisheries in the Economies of Pacific Island Countries and Territories*, Pacific Community, Noumea, New Caledonia, https://www.spc.int/sites/default/files/wordpresscontent/wp-content/uploads/2016/11/Gillett_16_Benefish-fisheries-in-economies-of-pacific-countries.pdf. [52]

Henriksson, P. et al. (2019), "Indonesian aquaculture futures - Identifying interventions for reducing environmental impacts", *Environmental Research Letters*, Vol. 14/12, p. 124062, http://dx.doi.org/10.1088/1748-9326/ab4b79. [62]

Himes-Cornell, A., S. Grose and L. Pendleton (2018), "Mangrove ecosystem service values and methodological approaches to valuation: Where do we stand?", *Frontiers in Marine Science*, Vol. 5, p. 376, http://dx.doi.org/10.3389/fmars.2018.00376. [38]

Hutniczak, B., C. Delpeuch and A. Leroy (2019), "Closing Gaps in National Regulations Against IUU Fishing", *OECD Food, Agriculture and Fisheries Papers*, No. 120, OECD Publishing, Paris, https://dx.doi.org/10.1787/9b86ba08-en. [50]

Huxham, M. (2013), *Mikoko Pamoja: Mangrove Conservation for Community Benefit*, Plan Vivo Foundation, Edinburgh, https://planvivo.org/docs/Mikoko-Pamoja-PDD_published.pdf. [42]

ITF (2018), "Reducing shipping greenhouse gas emissions: Lessons from port-based incentives", *International Transport Forum Policy Papers*, No. 48, OECD Publishing, Paris, https://dx.doi.org/10.1787/d3cecae7-en. [85]

Iyer, V. et al. (2018), *Finance Tools for Coral Reef Conservation: A Guide*, Conservation Finance Alliance, https://static1.squarespace.com/static/57e1f17b37c58156a98f1ee4/t/5c7d85219b747a7942c16e01/1551730017189/50+Reefs+Finance+Guide+FINAL-sm.pdf. [27]

Jambeck, J. et al. (2015), "Plastic waste inputs from land into the ocean", *Science*, Vol. 347/6223, pp. 768-771, http://dx.doi.org/10.1126/SCIENCE.1260352. [2]

Kenya Ministry of Environment and Forestry (n.d.), *Independent Trust Fund is Key to Securing Wildlife Assets, PS*, http://www.environment.go.ke/?p=3393 (accessed on July 2020). [30]

Kenya Wildlife Service (2019), *Conservation Fees*, http://www.kws.go.ke/sites/default/files/parksresorces%3A/KWS%20Conservation%20Fees%20Poster.pdf. [24]

KKP (2016), *Tentang UPTD KKP Raja Ampat (About Raja Ampat's UPTD KKP)*, http://kkpr4.net/en/index.php. [25]

Koelmans, A. et al. (2017), "Risks of plastic debris: Unravelling fact, opinion, perception, and belief", *Environmental Science & Technology*, Vol. 51/20, pp. 11513-11519, http://dx.doi.org/10.1021/acs.est.7b02219. [80]

Kollmuss, A. et al. (2008), *Making Sense of the Voluntary Carbon Market: A Comparison of Carbon Offset Standards*, World Wildlife Fund Germany, Frankfurt, https://mediamanager.sei.org/documents/Publications/SEI-Report-WWF-ComparisonCarbonOffset-08.pdf. [44]

La Dirección del Parque Nacional Galápagos [Galapagos National Park Directorate] (2018), *Informe Anuale 2017 [Annual Report 2017]*, `Ministerio del Ambiente, Ecuador, http://www.galapagos.gob.ec/wp-content/uploads/downloads/2018/02/informe_visitantes_anual_2017.pdf. [76]

Langenheim, J. (2017), *Indonesia pledges $1bn a year to curb ocean waste*, https://www.theguardian.com/environment/the-coral-triangle/2017/mar/02/indonesia-pledges-us1-billion-a-year-to-curb-ocean-waste. [83]

MacClellan, R. (2019), *Commentary: A call for Caribbean governments to tax cruise sector more and tax air passengers less*, https://www.caribbeanlifenews.com/stories/2019/7/2019-07-12-oped-caribbean-tourism-taxes-dilemma-cl.html. [71]

Macreadie, P. et al. (2019), "The future of Blue Carbon science", *Nature Communications*, Vol. 10/1, http://dx.doi.org/10.1038/s41467-019-11693-w. [46]

Menéndez, P. et al. (2018), "Valuing the protection services of mangroves at national scale: The Philippines", *Ecosystem Services*, Vol. 34, pp. 24-36, http://dx.doi.org/10.1016/j.ecoser.2018.09.005. [36]

Miedzinski, M., M. Mazzucato and P. Ekins (2019), "A framework for mission-oriented innovation policy roadmapping for the SDGs: The case of plastic-free oceans", *Working Paper Series*, No. 2019-03, UCL Institute for Innovation and Public Purpose, London, https://www.ucl.ac.uk/bartlett/public-purpose/wp2019-03. [81]

MSC (2014), *Marine Stewardship Council Global Impacts Report 2014: Monitoring and Evaluation (1999-2013)*, https://www.msc.org/docs/default-source/default-document-library/what-we-are-doing/global-impact-reports/msc_global-impacts-report-2014-web.pdf?sfvrsn=5f4e0f20_8. [16]

Mwamba, M. et al. (2018), *2017-2018 Plan Vivo Annual Report: Mikoko Pamoja*, Plan Vivo Foundation, Edinburgh, https://www.planvivo.org/docs/2017-2018_Mikoko-Pamoja-Annual-Report-Final-public_.pdf. [43]

Narayan, S. et al. (2019), *Valuing the Flood Risk Reduction Benefits of Florida's Mangroves*, The Nature Conservancy, Arlington, VA, http://www.conservationgateway.org/SiteAssets/Pages/floridamangroves/Mangrove_Report_digital_FINAL.pdf. [35]

National Council on Green Growth (2013), *National Strategic Plan on Green Growth (2013-2020)*, Royal Government of Cambogia, https://www.greengrowthknowledge.org/sites/default/files/downloads/policy-database/CAMBODIA%29%20National%20Strategic%20Plan%20on%20Green%20Growth%202013-2030.pdf. [1]

National Geographic Society (2020), *Prisine Seas: Expeditions (webpage)*, https://www.nationalgeographic.org/projects/pristine-seas/expeditions/ (accessed on 6 September 2019). [9]

Oceans, O. (ed.) (2017), *Financing spotlight: Blue Abadi, a $38-million trust fund to support MPAs in the Bird's Head region of Indonesia*, Open Communication for the Oceans, https://mpanews.openchannels.org/news/mpa-news/financing-spotlight-blue-abadi-38-million-trust-fund-support-mpas-birds-head-region (accessed on July 2020). [29]

OECD (2020), *Policy Instruments for the Environment (database)*, http://oe.cd/pine. [15]

OECD (2020), "Rethinking tourism success for sustainable growth", in *OECD Tourism Trends and Policies 2020*, OECD Publishing, Paris, https://dx.doi.org/10.1787/82b46508-en. [64]

OECD (2019), *Fisheries Support Estimate Database*, http://oe.cd/FSE (accessed on 2019). [98]

OECD (2019), *OECD Environment Statistics - Biodiversity: Protected Areas (database)*, http://dx.doi.org/10.1787/5fa661ce-en (accessed on 15 May 2019). [8]

OECD (2019), *Responding to Rising Seas: OECD Country Approaches to Tackling Coastal Risks*, OECD Publishing, Paris, https://dx.doi.org/10.1787/9789264312487-en. [33]

OECD (2019), *Rethinking Innovation for a Sustainable Ocean Economy*, OECD Publishing, Paris, https://dx.doi.org/10.1787/9789264311053-en. [86]

OECD (2018), "Improving plastics management: Trends, policy responses, and the role of international co-operation and trade", *OECD Environment Policy Papers*, No. 12, OECD Publishing, Paris, https://dx.doi.org/10.1787/c5f7c448-en. [77]

OECD (2018), *OECD Tourism Trends and Policies 2018*, OECD Publishing, Paris, https://dx.doi.org/10.1787/tour-2018-en. [66]

OECD (2017), *Marine Protected Areas: Economics, Management and Effective Policy Mixes*, OECD Publishing, Paris, https://doi.org/10.1787/9789264276208-en (accessed on 1 May 2018). [4]

OECD (2017), "Support to fisheries: Levels and impacts", *OECD Food, Agriculture and Fisheries Papers*, No. 103, OECD Publishing, Paris, https://dx.doi.org/10.1787/00287855-en. [54]

OECD (2017), *The Political Economy of Biodiversity Policy Reform*, OECD Publishing, Paris, http://dx.doi.org/10.1787/9789264269545-en. [32]

OECD (2016), *Biodiversity Offsets: Effective Design and Implementation*, OECD Publishing, Paris, https://dx.doi.org/10.1787/9789264222519-en. [14]

OECD (2014), *OECD Tourism Trends and Policies 2014*, OECD Publishing, Paris, https://dx.doi.org/10.1787/tour-2014-en. [69]

OECD (2011), *Fisheries and Aquaculture Certification*, OECD Publishing, Paris, https://dx.doi.org/10.1787/9789264119680-en. [17]

OECD (2007), *Instrument Mixes for Environmental Policy*, OECD Publishing, Paris, https://dx.doi.org/10.1787/9789264018419-en. [3]

OECD/FAO (2019), *OECD-FAO Agricultural Outlook 2019-2028*, OECD Publishing, Paris/Food and Agriculture Organization of the United Nations, Rome, https://dx.doi.org/10.1787/agr_outlook-2019-en. [55]

Office of Prime Minister of Thailand (2017), *Twelfth National Economic and Social Development Plans (2017-2021)*, https://www.greengrowthknowledge.org/sites/default/files/downloads/policy-database/THAILAND%29%20The%20Twelfth%20National%20Economic%20and%20Social%20Development%20Plan%20%282017-2021%29.pdf. [61]

Ollivaud, P. and P. Haxton (2019), "Making the most of tourism in Indonesia to promote sustainable regional development", *OECD Economics Department Working Papers*, No. 1535, OECD Publishing, Paris, https://dx.doi.org/10.1787/c73325d9-en. [73]

Organisation of Eastern Caribbean States (2019), *OECS to establish Green-Blue Economy Strategy and Action Plan with CANARI*, https://pressroom.oecs.org/oecs-to-establish-green-blue-economy-strategy-and-action-plan-with-canari. [96]

PLOCAN (2020), *EcoMarPort, Transferencia tecnológica y Eco-innovación para la gestión ambiental y Marina en zonas Portuarias de la Macaronesia*, https://www.ecomarport.eu/. [91]

Pollnac, R. et al. (2010), "Marine reserves as linked social-ecological systems", *Proceedings of the National Academy of Sciences*, Vol. 107/43, pp. 18262-18265, http://dx.doi.org/10.1073/pnas.0908266107. [26]

Pörtner, H. et al. (eds.) (2019), *Summary for policymakers*, Intergovernmental Panel on Climate Change (IPCC), Geneva. [34]

Reyes, C. et al. (2018), *The Boracay Closure: Socioeconomic Consequences and Resilience Management*, Philippine Institute for Development Studies, Quezon City, Philippines, https://pidswebs.pids.gov.ph/CDN/PUBLICATIONS/pidsdps1837.pdf (accessed on 14 September 2019). [67]

Roberts, J. and A. Ali (2018), *The Blue Economy and Small States*, Commonwealth Secretariat, London, http://dx.doi.org/978-1-84859-950-5. [90]

Rosales, R. et al. (2017), "Value chain analysis and small-scale fisheries management", *Marine Policy*, Vol. 83, pp. 11-21, http://dx.doi.org/10.1016/j.marpol.2017.05.023. [56]

Sampantamit, T. et al. (2020), "Aquaculture production and its environmental sustainability in Thailand: Challenges and potential solutions", *Sustainability*, Vol. 12/5, p. 2010, http://dx.doi.org/10.3390/su12052010. [60]

Sargasso Sea Commission (2019), *Sargasso Sea Commission (webpage)*, http://www.sargassoseacommission.org/index.php (accessed on 22 July 2019). [97]

Scott, D., C. Hall and S. Gössling (2019), "Global tourism vulnerability to climate change", *Annals of Tourism Research*, Vol. 77, pp. 49-61, http://dx.doi.org/10.1016/j.annals.2019.05.007. [65]

SeyCCAT (2019), *SeyCCAT - About Us (web page)*, https://seyccat.org/about-us/. [7]

South Africa Department of Environmental Affairs (2019), *Oceans Economy Summary Progress Report February 2019*, South Africa Department of Environmental Affairs, Pretoria, https://www.environment.gov.za/sites/default/files/reports/oceanseconomy_summaryprogressreport2019.pdf (accessed on 23 July 2019). [51]

Sud, M. (2020), "Managing the biodiversity impacts of fertiliser and pesticide use: Overview and insights from trends and policies across selected OECD countries", *OECD Environment Working Papers*, No. 155, OECD Publishing, Paris, https://dx.doi.org/10.1787/63942249-en. [84]

The Nature Conservancy (2019), *Seychelles expands marine protections to 26% - Achieves milestone 2 of the award-winning debt conversion*, https://www.nature.org/en-us/explore/newsroom/seychelles-expands-marine-protections/. [6]

The Nature Conservancy (2010), *Coral Triangle MCA Feasibility Analysis – Final Interim Findings Public Version, V.2*, https://www.mpaaction.org/sites/default/files/MCA_Feasibility_Analysis_Coral_Triangle_Indonesia_Public_Version_Final_2010_Oct_V2_ENGLISH.pdf. [75]

Thompson, A. et al. (2014), *Tourism Concessions in Protected Areas: Guidlines for Managers*, United Nations Development Programme, New York, https://www.undp.org/content/undp/en/home/librarypage/environment-energy/ecosystems_and_biodiversity/tourism-concessions-in-protected-natural-areas.html. [23]

Thrush, S. (ed.) (2012), "Estimating global 'blue carbon' emissions from conversion and degradation of vegetated coastal ecosystems", *PLoS ONE*, Vol. 7/9, p. e43542, http://dx.doi.org/10.1371/journal.pone.0043542. [39]

U.S. Fish and Wildlife Service (2017), "U.S. Fish and Wildlife Service statement from Bryan Arroyo, Assistant Director for International Affairs, on the expansion of Gabon's Marine Protected Area Network", https://www.fws.gov/international/wildlife-without-borders/africa/statement-gabon-mpa-expansion.html (accessed on 22 July 2019). [10]

UN (2019), *Sustainable Development Goals Knowledge Platform (webpage)*, https://sustainabledevelopment.un.org/sdg14. [88]

UN (2018), *Report of the Secretary-General: Sustainable Development of the Caribbean Sea for Present and Future Generations*, United Nations General Assembly, New York, https://www.un.org/ga/search/view_doc.asp?symbol=A/73/225&Lang=E. [95]

UN Environment (2019), *Naples Ministerial Declaration: Mediterranean Action Plan*, United Nations Environment Programme (UNEP), Nairobi, https://www.msp-platform.eu/sites/default/files/naples_declaration_0.pdf. [87]

UN Environment (2018), *Single-use Plastics: A Roadmap for Sustainability*, United Nations Environment Programme, Nairobi, https://wedocs.unep.org/bitstream/handle/20.500.11822/25496/singleUsePlastic_sustainability.pdf. [82]

UNCTAD (2019), *World Investment Report 2019: Special Economic Zones*, United Nations Conference on Trade and Development (UNCTAD), Geneva, https://unctad.org/en/PublicationsLibrary/wir2019_en.pdf. [72]

UNDP (2019), *The Biodiversity Finance Initiative - Seychelles: Finance Plan for Biodiversity Conservation 2019-2023*, United Nations Development Programme (UNDP), New York, https://www.biodiversityfinance.net/sites/default/files/content/knowledge_products/BIOFIN%20Finance%20Plan%20-%20%2015th%20February%202019.pdf. [70]

UNECA (2016), *Africa's Blue Economy: A Policy Handbook*, United Nations Economic Commission for Africa (UNECA), Addis Ababa, https://www.uneca.org/publications/africas-blue-economy-policy-handbook. [94]

VCS (2019), *Coastal Carbon Corridor: Mangrove Restoration and Coastal Greenbelt Protection in the East Coast of Aceh and North Sumatra Province, Indonesia*, Verified Carbon Standard. [45]

Waldron, A. et al. (2017), "Reductions in global biodiversity loss predicted from conservation spending", *Nature*, Vol. 55, pp. 364-367, http://dx.doi.org/10.1038/nature24295. [22]

Watson, J. et al. (2014), "The performance and potential of protected areas", *Nature*, Vol. 515/7525, pp. 67-73, http://dx.doi.org/10.1038/nature13947. [20]

World Bank (2017), *The Sunken Billions Revisited: Progress and Challenges in Global Marine Fisheries*, World Bank, Washington, DC, http://dx.doi.org/10.1596/978-1-4648-0919-4. [48]

World Economic Forum (2016), *The New Plastics Economy: Rethinking the Future of Plastics*, http://www3.weforum.org/docs/WEF_The_New_Plastics_Economy.pdf. [79]

World Tourism Organization (2019), *International Tourism Highlights: 2019 Edition*, http://dx.doi.org/10.18111/9789284421152. [74]

World Tourism Organization (2019), *World Tourism Barometer*. [19]

World Tourism Organization (2002), *Tourism and Poverty Alleviation*. [68]

World Travel & Tourism Council (2019), *Economic Impact of Global Wildlife Tourism*, https://travesiasdigital.com/wp-content/uploads/2019/08/The-Economic-Impact-of-Global-Wildlife-Tourism-Final-19.pdf. [18]

Wunder, S. (2015), "Revisiting the concept of payments for environmental services", *Ecological Economics*, Vol. 117, pp. 234-243, http://dx.doi.org/10.1016/j.ecolecon.2014.08.016. [13]

Wylie, L., A. Sutton-Grier and A. Moore (2016), "Keys to successful blue carbon projects: Lessons learned from global case studies", *Marine Policy*, Vol. 65, pp. 76-84, http://dx.doi.org/10.1016/j.marpol.2015.12.020. [37]

Annex 3.A. Examples of policy instruments to address different pressures on the ocean

Annex Table 3.A.1. Examples of regulatory policy instruments to address different pressures on the ocean

Regulatory instruments	Pressure				
	Overfishing	Pollution	Habitat destruction	Invasive alien species	Climate change
Marine spatial planning (spatial restrictions for specific activities)	X	X	X	X	X
Marine protected areas	X	x	X		
Temporal restrictions (seasonal, temporary closures)	X	X	X	X	X
Total allowable catch	X				
Individual catch quotas	X				
Territorial use rights	X				
Property rights	X				
Effort quotas (limits on the number of days at sea)	X				
Fishing standards	X				
Fishing licenses	X				
Gear restrictions	X		X		
By-catch restrictions	X				
Discard restrictions/bans	X				
Landing limits (restrictions on fish quantities and size)	X				
Vessel restrictions (number, size, horsepower)	X	X			
Ship construction standards	X				
Specification of "best available technology" or "best environmental practice" for fishing	X	X	X	X	X
Planning requirements (i.e. environmental impact assessments and emergency response plans)		X	X	X	X
Standards (e.g. pollution, emissions, construction)		X	X	X	X
Emission permits		X			X
Restrictions on mineral extraction		X			
Restrictions on ballast water discharges		X		X	
Restrictions on volume and concentration of discharged pollutants from onshore and offshore		X			
Limitation on oil, gas and other mining operations		X	X		X
Limitation on number of freight and cruise ships operating		X		X	
Restrictions on tourism operations		X	X		

Annex Table 3.A.2. Examples of economic and information and voluntary policy instruments to address different pressures on the ocean

Economic instruments and information and voluntary approaches	Pressure				
	Overfishing	Pollution	Habitat destruction	Invasive alien species	Climate change
Economic instruments					
Individually transferable quotas	X				
Resource tax	X				
User fees	X	X	X	X	
R&D subsidies	X	X	X	X	X
Non-compliance fees/penalties	X	X	X	X	X
Insurance measures	X				
Removal or reform of harmful subsidies	X	X	X	X	X
Buy-back and decommissioning schemes	X	X	X	X	X
Taxes on fertilisers and pesticides (inputs)		X			
Pollution taxes or emissions trading schemes		X	X		X
Payments for ecosystem services	X	X	X		
Information and voluntary approaches					
Certification and eco-labelling	X	X	X		X
Industry codes of practice	X	X	X	X	X
Marine charts, navigation aids, other marine services	X	X	X	X	X
Awareness campaigns and education	X	X	X	X	X

Source: OECD (2017[4]), *Marine Protected Areas: Economics, Management and Effective Policy Mixes*, https://doi.org/10.1787/9789264276208-en.

Notes

[1] The Caribbean Challenge Initiative was launched at the ninth meeting of the Conference of the Parties to the Convention on Biological Diversity. In addition to the second goal shown here, it set out the goal to effectively conserve and manage at least 20% of the marine and coastal environment by 2020..

[2] On the compliance markets, a buyer is obliged to purchase carbon offsets to meet emissions reduction targets under a particular treaty or mechanism, internationally or domestically (such as the Kyoto protocol or the EU Emissions Trading System). Voluntary markets serve entities that purchase carbon offsets voluntarily to be more sustainable.

[3] Where financial revenues from tourism activities are remitted abroad.

[4] For more information, see https://www.tourismauthority.go.ke/index.php/tourism-support-services/custom-duty-exemption.

[5] Entrance fees contributed the largest amount to the operating budget, at USD 11.4 million.

4 Development co-operation for sustainable ocean economies

This chapter, along with Chapter 5, examines how development co-operation can help create a global ocean economy that is sustainable and responds to the needs of developing countries and the world's most vulnerable. It provides the first official tracking of official development assistance (ODA) for sustainable ocean economies to highlight the scope, nature and trends of international support for sustainable ocean economies in developing countries. The chapter also maps the range of ODA projects in support of sustainable ocean economies to help foster a common understanding of the multiple dimensions of sustainability and how they vary across ocean-based sectors. The chapter concludes with suggestions to enhance the impact of development co-operation in support of sustainable ocean economies.

Development co-operation has a critical role in supporting developing countries to harness the benefits of sustainable ocean economies

Global urgency and ambition for conserving and sustainably using the ocean are rising. If managed sustainably, the ocean has the capacity to regenerate, be more productive and support more equitable societies. However, not all countries are in a position to undertake the bold, forceful actions required to harness the benefits of sustainable ocean economies. International co-operation and fair global arrangements have a key role to play in ensuring that the expansion of the global ocean economy is guided by institutional arrangements, policies and financial flows that are aligned with the imperative of sustainability and with the needs of developing countries and the world's poorest and most vulnerable people. Development co-operation has also significant role to play to facilitate developing countries' access to the science, evidence, innovations, and financial resources needed to transform both emerging and existing ocean-based sectors into catalysts for long-term, inclusive and sustainable development.

The impacts of the COVID-19 pandemic make the need for resolute ocean action from the development community all the more pressing. Entire ocean-based sectors such as tourism, sea transport and many others are being disrupted, with major economic consequences for developing countries. While it is too early to know the full impact of the pandemic, the needs of the post-crisis recovery will certainly vary across countries and recovery efforts will have to be country-led. Development co-operation will need to be in line with countries' national priorities and strategies for the emergency response and the recovery. It will need to concentrate not only on helping to address the health emergency, but also on 'building back bluer' fostering a recovery that puts ocean-based sectors solidly on a track of environmental and social sustainability. Official development assistance (ODA) and other resources – public, private, international and domestic – will likely come under increasing pressure. To maintain their focus on long-term sustainability, it will be essential to further identify priorities and approaches that can maximise the socio-economic and environmental benefits of ODA and other interventions.

This chapter presents original evidence on the volume, scope and nature of development co-operation for the ocean economy, highlighting the extent to which it incorporates sustainability and the approaches through which it is provided. The chapter provides the first quantification and analysis of ocean-related ODA, developed specifically for this report. The chapter also builds on the findings of the survey of members of the OECD Development Assistance Committee (DAC) conducted for to inform this report, the OECD Survey on DAC Members' Policies and Practices in Support of the Sustainable Ocean Economy (OECD Survey), and the Sustainable Ocean Economy Country Diagnostics conducted to inform this report. Figure 4.1 and Figure 4.2 illustrate key facts on ocean-relevant ODA, including its evolution over time and how it is distributed between regions, sectors and Sustainable Development Goal (SDG) 14 targets.

Figure 4.1. Key figures on ODA for the ocean economy and ODA for the sustainable ocean economy (2013-18)

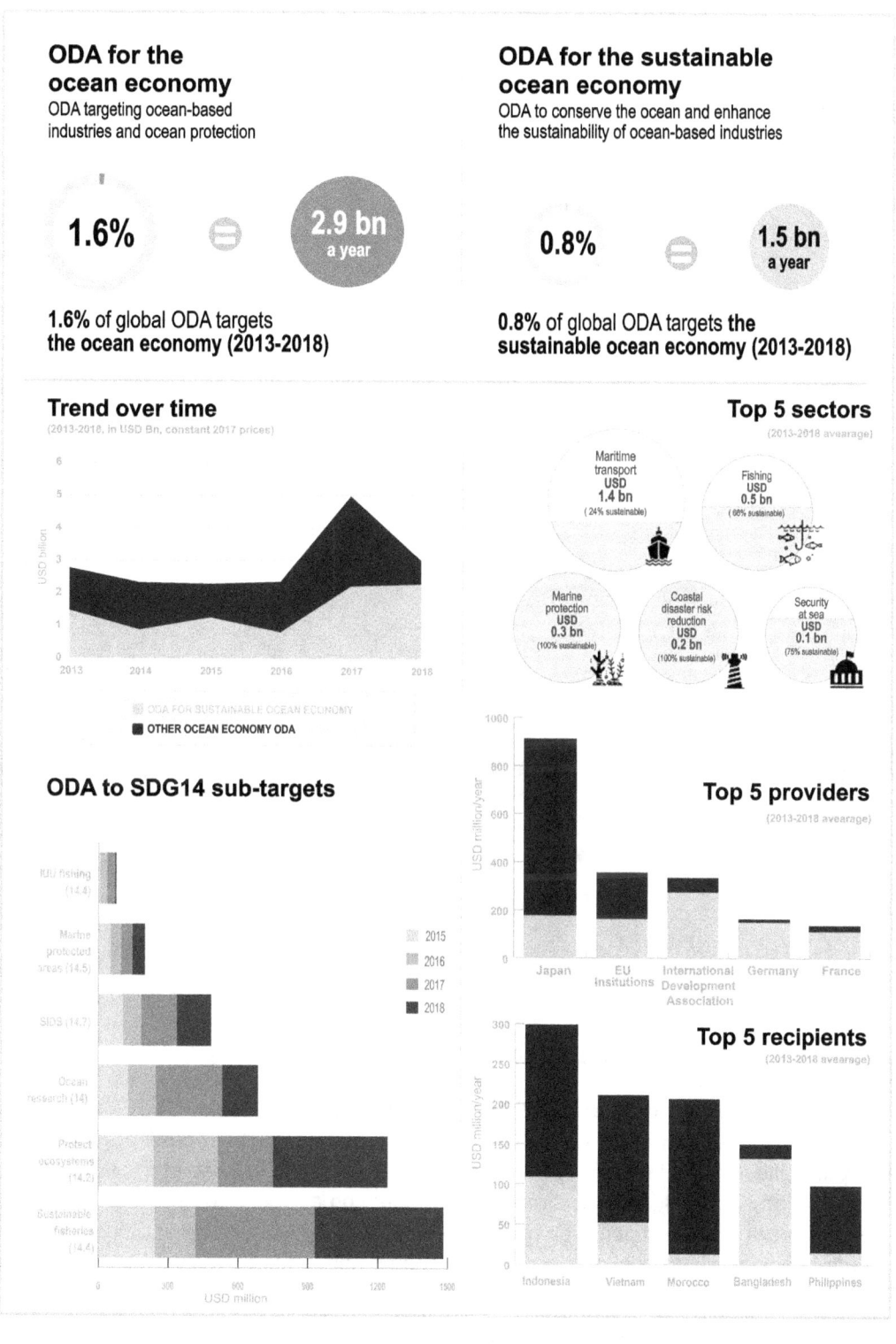

Source: Authors' calculations based on OECD (2020[1]), Creditor Reporting System (database), https://stats.oecd.org/Index.aspx?DataSetCode=crs1

StatLink https://doi.org/10.1787/888934159430

Figure 4.2. Key figures on ODA to reduce ocean pollution from land (2013-18)

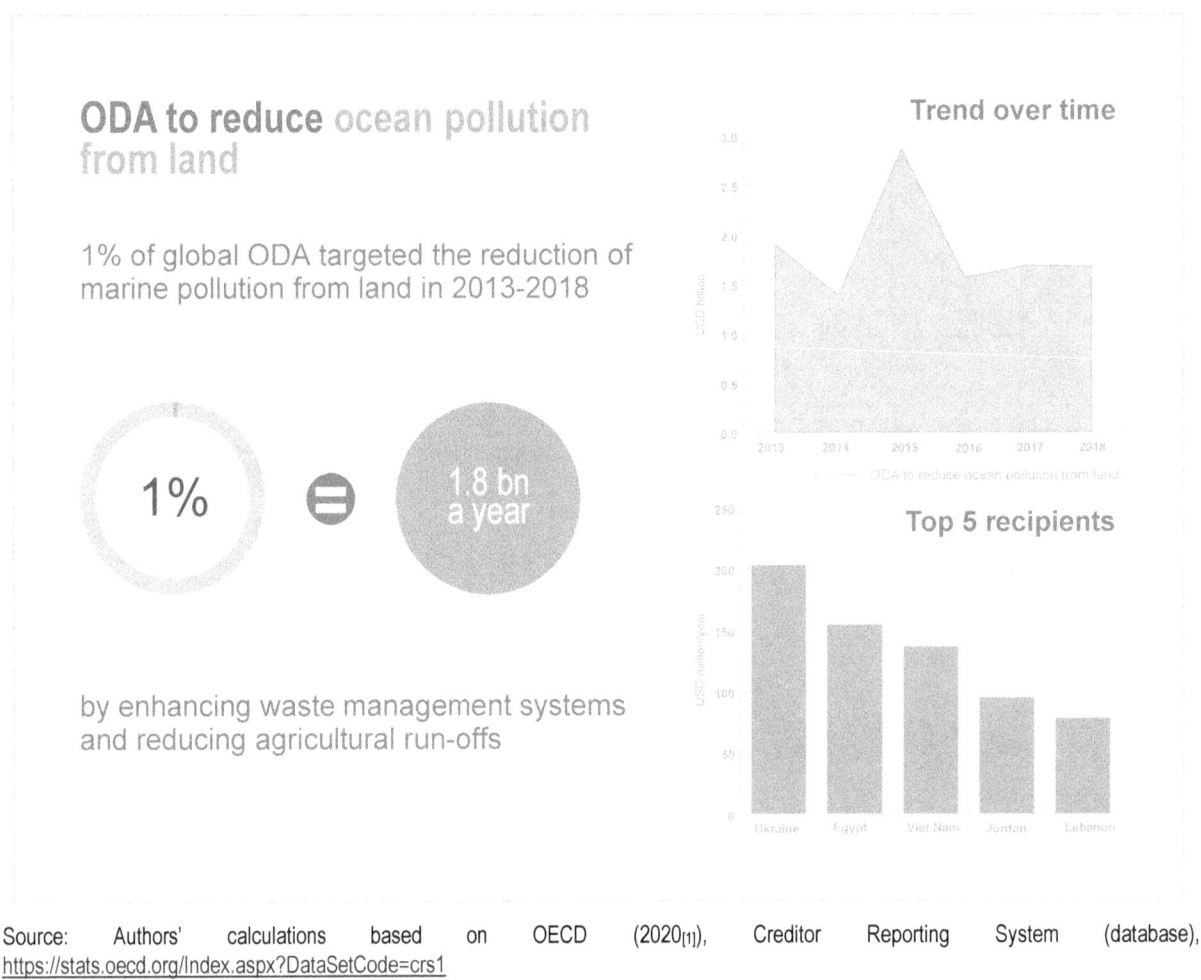

Source: Authors' calculations based on OECD (2020[1]), Creditor Reporting System (database), https://stats.oecd.org/Index.aspx?DataSetCode=crs1

StatLink https://doi.org/10.1787/888934159449

Growing international attention but no common understanding of the sustainable ocean economy to guide development co-operation

The international community's engagement on ocean is growing

Across the globe, an increasing number of initiatives are focusing on the ocean. The 2030 Agenda and the SDGs, especially SDG 14, as well as domestic pressure to address global ocean pollution and the new economic opportunities arising from the ocean economy are all driving this rise in a focus on ocean action (OECD Survey). Many new initiatives and much of the attention are centred on ocean plastics, an issue that has become more of a focal point for concern about the marine environment than the more radical changes to our behavioural, political and economic systems needed to achieve a healthy and productive ocean (Stafford and Jones, 2019[2]).

To foster sustainable ocean economies, countries are developing new policies, forging new alliances and creating new funds. Japan recently leveraged its Presidency of the G20 to achieve the endorsement of the Osaka Blue Ocean Vision, which aims to reduce additional pollution from marine plastic litter to zero by 2050. Canada put oceans at the centre of its G7 presidency in 2018, launching the Charlevoix Blueprint for Healthy Oceans, Seas and Resilient Coastal Communities and advancing the Oceans Plastics Charter. To date, 26 governments and 69 businesses and organisations have signed the Charter. Another initiative is the High Level Panel for a Sustainable Ocean Economy, created by leaders of 14 countries[1] from the global North and South and co-chaired by Norway and Palau as a collaborative effort to advance a sustainable ocean economy. In addition, the United Kingdom is establishing a GBR 500million (United Kingdom pound) Blue Planet Fund to help eligible countries protect their marine resources from key, human-generated threats including climate change and habitat loss.

Multilateral institutions that traditionally did not focus on the ocean are creating dedicated programmes and partnerships. In 2018, a group of donor countries established PROBLUE, a USD 150-million multi-donor trust fund that is hosted at the World Bank. Its four main priorities are to promote sustainable fisheries and aquaculture, curb marine pollution, enhance the sustainability of key sectors in the ocean economy, and support governments to build the necessary capacity to manage marine resources. In 2019, the Asian Development Bank (ADB) launched the Action Plan for Healthy Oceans and Sustainable Blue Economies for the Asia and Pacific region, a USD 5-billion plan targeting sustainable tourism and fisheries, the protection of marine and coastal ecosystems, reduction of land-based pollution, and enhancement of the sustainability of port and coastal infrastructure. In 2019, the French Development Agency, KfW and the European Investment Bank established the Clean Oceans Initiative to mobilise USD 2 billion over a five-year horizon to support countries, cities and businesses to manage and process their waste more efficiently and reduce ocean waste.

Philanthropies are demonstrating an increasing interest in ocean issues, recently investing in large marine conservation projects and stepping up support for sustainable fisheries and to address marine litter. The OECD estimates that in 2018, philanthropy provided USD 200 million for the ocean. The largest of these donors were the Oak Foundation (44% of the total), the David and Lucile Packard Foundation (15%), the Dutch Postcode Lottery (12%), and the MAVA Foundation (10%). From a geographical perspective, the most targeted regions are South of Sahara (51% of regionally allocable aid) and Far East Asia (19%).

No common definitions and dedicated cooperation tools for development cooperation in support of sustainable ocean economies Despite the international community's growing interest in ocean matters, there remains a lack of common understandings, definitions and principles with which to align co-operation efforts and ensure that the development community is moving effectively towards the same targets. One of the reasons is that the sustainable ocean economy concept is fairly new. While several development co-operation providers have long track records of support for marine protection or specific ocean-based sectors such as fisheries, few have a more holistic understanding of the ocean economy and have started to consider ocean-relevant interventions through an integrated, cross-sectoral approach and a sustainability lens.

According to the OECD Survey responses, no DAC member has adopted an official definition of the sustainable ocean economy to guide its development co-operation efforts. Some DAC members employ working definitions of the sustainable ocean economy. All of these appear grounded in the idea that there is a need to reconcile economic opportunities with safeguarding the long-term health of the marine environment, although the understanding of what this means in practice varies widely. This is reflected in the sectors identified by survey respondents as being part of a sustainable ocean economy. For instance, four respondents list offshore oil and gas and seabed mining, as these industries represent a large share of the ocean economy gross domestic product, while seven only industries with more limited environmental impact and greater potential to become sustainable.

> *"Where there are explicit trade-offs between conservation (or climate action) objectives, and economic development objectives, often the conservation/climate objective can get lost. Getting the balance right across those objectives is a crucial feature of sustainable ocean economies and should be reinforced at every opportunity. The shared experience of OECD members and international institutions should explicitly assist countries in addressing these trade-offs"* – A survey respondent

Most respondents identify climate change mitigation, adaptation and resilience as integral parts of the environmental sustainability aspect of sustainable ocean economies. Only three respondents instead explicitly mentioned social sustainability – i.e. improving livelihoods and well-being, creating decent jobs, and fostering socio-economic inclusion – in their working definitions. Thus, while many countries recognise the importance of ocean industries to promote the livelihoods of coastal populations, very few adopt a definition of sustainability that goes beyond the environmental dimension and takes into account social sustainability in line with the 2030 Agenda's three dimensions of sustainability. Most respondents (11 of 13) see sustainable fisheries as the sustainable ocean economy sector with the largest potential in developing countries, although fewer also include sustainable tourism (6 of 13) or aquaculture (5 of 13). The importance of marine renewable energy is highlighted by 4 of 13 respondents, although this was the least targeted ocean sector in terms of ODA until 2017.

Most providers of development co-operation lack dedicated development co-operation strategies and implementation and monitoring tools for promoting sustainable ocean economies. Only four countries to date have either developed a standalone strategy on development co-operation in support of the ocean (Iceland and Norway) or are in the process of developing such a strategy (Ireland and France). All other bilateral donors tend to address issues related to the ocean through a thematic or sectoral approach. For instance, most donors have a specific strategy for fishery support and some also carry out targeted impact evaluations to assess the effectiveness of their funding in this sector. As a result, ocean issues are addressed in a fairly fragmented manner rather than as part of integrated ocean management.

Towards common definitions: tracking ODA trends for sustainable ocean economies

Effective development co-operation in support of sustainable ocean economies requires a common understanding, definitions and principles regarding what *sustainable* ODA interventions in the ocean economy space consist of. Such definitions and principles are needed to provide accountability and visibility for ODA spending and to monitor how much ODA is contributing to achieve sustainable ocean economies, global priorities and targets on the ocean, and SDG 14. Common definitions also can foster shared learning across providers and help identify good practices and scalable approaches that can increase the effectiveness of ODA interventions for the ocean economy. Further, evidence on global finance for the ocean - from private, public, domestic and international sources - is still scarce and scattered. Tracking ODA for the sustainable ocean economy contributes to filling this gap and to enhancing transparency on the range of global financial flows for the ocean.

For these reasons and as part of the Sustainable Ocean for All initiative, the OECD has started to quantify and track global development finance for the ocean, detailing its scope, sources and destinations and providing estimates of the share of this finance that is sustainable. A specific methodology was developed for this purpose that lays the foundations for a common understanding of what defines ODA interventions in support of sustainable ocean economies. This methodology and the key indicators to track ODA for the ocean economy are briefly explained in Box 4.1 and more extensively discussed in Annex 4.A. The remainder of this chapter builds on an analysis of ODA flows for the sustainable ocean economy as identified via this methodology.

Box 4.1. Indicators and definitions used in the analysis of ocean-related ODA

Evidence on global finance for the ocean from its various sources – private, public, domestic and international – remains scarce and scattered. It is currently not possible to have a comprehensive view of how much finance reaches ocean-based sectors and what percentage of this can be considered sustainable. To contribute to fill this gap and as part of the Sustainable Ocean for All initiative, the OECD has begun to quantify and track global development finance for the ocean, detailing its scope, sources and destinations and providing estimates of the share that is sustainable. Development finance estimates are also produced for funding towards land-based activities that reduce negative impacts on the ocean (e.g. waste management and water treatment).

The tracking of ocean-relevant ODA is based on the statistical data made available by the OECD DAC Creditor Reporting System (CRS), which provides a unique and comprehensive source of activity-level development finance. As there is no marker or immediate way to retrieve data on ODA for the ocean,[2] a specific methodology was developed to generate the first official estimates of ocean-relevant ODA. These provide the quantitative base for the analysis in this chapter and inform the report overall.

Figure 4.3. Three key indicators to track ocean-relevant ODA

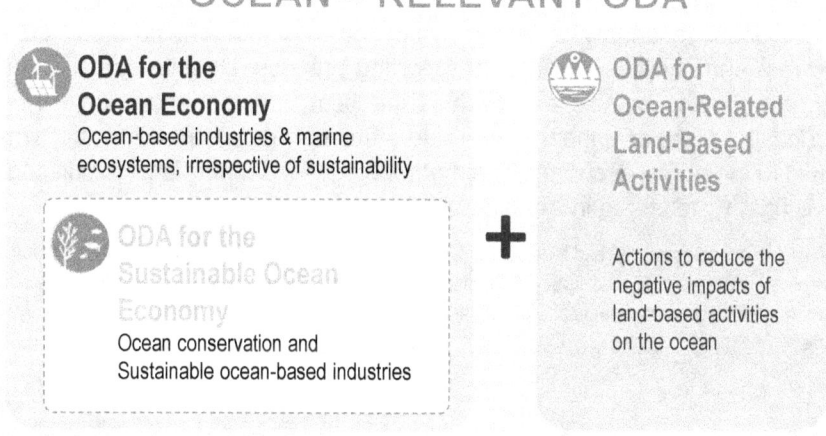

Source: Authors

Ocean-relevant ODA estimates are organised around three key indicators (Figure 4.3):

- **ODA for the ocean economy:** Also referred to as ocean ODA or ODA for the ocean, this is ODA in support of ocean-based industries and marine ecosystems, irrespective of whether the support explicitly takes sustainability considerations into account. For instance, fisheries projects with no specific focus on sustainable development would be included, as would projects in support of offshore oil and gas.

- **ODA for the sustainable ocean economy**: This is a subset of ODA for the ocean economy. It identifies ocean conservation activities as well as support for ocean-based industries that integrates sustainability concerns. For instance, projects in support of mangroves restoration would be captured, as would sustainable coastal tourism and sustainable fisheries projects.

- **ODA for reducing ocean pollution from land:** This capture land-based activities that reduce negative impacts and/or have a positive impact on ocean, such as water treatment and waste management projects. This indicator is included in recognition of the strong interrelation between land-based and marine activities and the fact that most ocean pollution originates from land-based activities.

The indicators draw on definitions of the ocean economy and the sustainable ocean economy.

Ocean economy: The OECD (2016[3]) defines the ocean economy as comprising ocean-based industries that depend, either directly or indirectly, on ocean resources (Annex 4.A). These include traditionally exploited marine resources, including living resources (fisheries) and non-living resources (oil, gas and marine manufacturing and construction. The definition also covers the use of oceans for tourism, education and shipping and ocean-based industries that have recently emerged due to advancements in science and technology such as offshore wind, tidal and wave energy; marine aquaculture; seabed mining for metals and minerals; and marine biotechnology. The definition of ODA for the ocean economy relies on this ocean economy definition.

Sustainable ocean economy: The sustainable ocean economy emphasises the sustainable use and conservation of natural resources in the world's oceans, seas and coastal areas, in line with the 2030 Agenda and the SDGs pertaining to the ocean. The ODA for the sustainable ocean economy indicator thus captures ODA in support of ocean-based and coastal economic activities that explicitly integrate sustainability such as sustainable fisheries (SDG target 14.4); specific activities to enhance ocean health (targets 14.2 and 14.3); activities to conserve marine and coastal ecosystems (targets 14.2 and 14.5); and activities to increase resilience and climate action. The identification of activities belonging to the sustainable ocean economy is largely based on development partners' self-assessment of sustainable and resilient activities in the data reporting process. Despite some limitation inherent to keyword search methodologies, this approach is the best approximation currently available and it provides a solid basis to further enhance the understanding and measurement of what sustainability means with regard to the ocean economy. The methodology for defining and calculating ocean-relevant ODA indicators is briefly presented in Annex 4. A.

Note: The terms sustainable ocean economy" and blue economy are sometimes used interchangeably to identify economies where ocean resources and marine ecosystems are used sustainably and conserved. The terms ODA and concessional finance are used interchangeably. Figures in this report refer to ODA commitments. The definition of ODA is available at http://www.oecd.org/development/financing-sustainable-development/development-finance standards/officialdevelopmentassistancedefinitionandcoverage.htm:.

The volume of ODA for the ocean is growing, but remains small and only partially focused on sustainability

The original ODA estimates produced for this report suggest that, over the 2013-18 period, an average of USD 3 billion of ODA a year was allocated for the ocean economy, equivalent to 1.6% of global ODA over the same period. Ocean economy ODA flows spiked in 2017 (Figure 4.4) and their growth rate in 2013-18 outpaced that of global ODA (+57% vs +17%).

Despite this acceleration, ODA for the ocean economy accounts for only a fraction of both global ODA (1.6%) and ODA to all coastal and island countries (1.9%) in 2013-18. More importantly, only about half (51%) of ODA for the ocean economy contributed to sustainable ocean economies, i.e. to ocean conservation and sustainable coastal and ocean economic activities. ODA for the sustainable ocean economy totaled USD 2.3 billion in 2018 and USD 1.5 billion on average a year in 2013-18, equivalent to 0.8% of global ODA in 2013-18 and 1% of all ODA to coastal and island countries.

In 2013-18, development partners have also provided concessional finance for land-based activities that reduce negative impacts on the ocean including projects to enhance waste management systems and for wastewater treatment. This support amounted to USD 1.7 billion of ODA in 2018, in line with the annual average in 2013-18.

Figure 4.4. Ocean-relevant ODA peaked in 2017 but still only accounts for a fraction of global ODA

ODA commitments in constant 2017 prices, 2013-18

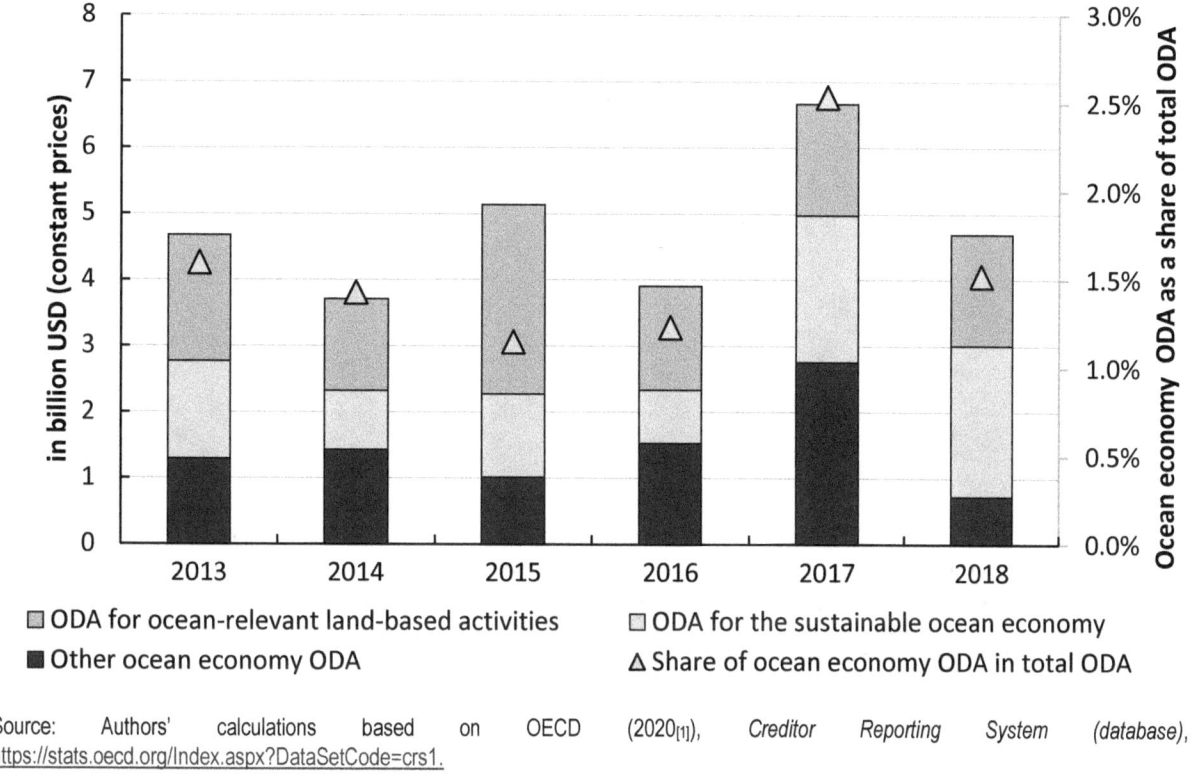

Source: Authors' calculations based on OECD (2020[1]), Creditor Reporting System (database), https://stats.oecd.org/Index.aspx?DataSetCode=crs1.

StatLink ᗢ https://doi.org/10.1787/888934159468

The low level of concessional finance for the conservation and sustainable use of the ocean can be understood by comparing it to concessional finance for other SDGs. Not all SDGs require the same level of financing and, as illustrated in Table 1.1. in Chapter 1, their interconnectedness determines that targeting some SDGs can indirectly contribute to achieving others (UN, 2015[4]; International Council for Science, 2017[5]). However, across SDGs, SDG14 is largely under-prioritised. Figure 4.5 shows it is among the least targeted of the SDGs, both by official development finance and private philanthropy for development.

Figure 4.5. SDG 14 is among the least funded SDGs by both Official Development Assistance and philanthropic development funding

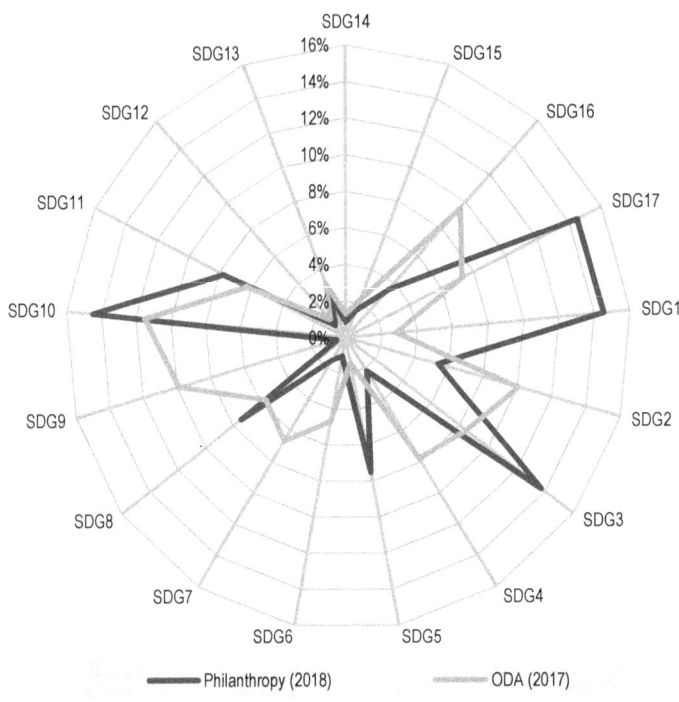

Note: The relative share for each SDG is calculated based on the sum of 2017 ODA commitments towards each SDG, as reported on the OECD SDG Financing Lab website, https://sdg-financing-lab.oecd.org. Philanthropy data are 2018 commitments as reported in the Creditor Reporting System database, https://stats.oecd.org/Index.aspx?DataSetCode=crs1.
Source: Authors' calculations based on OECD (2020[1]), *Creditor Reporting System (database)*, https://stats.oecd.org/Index.aspx?DataSetCode=crs1.

StatLink https://doi.org/10.1787/888934159487

Box 4.2. Mainstreaming gender equality in the sustainable ocean economy

SDG 5 (gender equality and women's empowerment) and SDG 14 are inextricably linked

Women are essential to the sustainable ocean economy. Roughly 15% of harvesting jobs and 90% of fish processing jobs are done by women (Siles et al., 2019[6]). However, their role in many ocean-based sectors is unrecognised and informal, and they are marginalised, under-represented and effectively excluded from many economic opportunities and decision-making positions. As leaders, citizens, resources users, educators and scientists, women need to be empowered to engage more in ocean conservation and sustainable use. To bring this about, affirmative actions are required to mainstream gender equality best practices into every sphere of ocean-based activities.

The role of women in the fishery sector illustrates this state of affairs. In the fisheries industry, women's role along the value chain is often undervalued and under-recognised, making it difficult even to collect information about their actual involvement in the industry. This in turn gives rise to their under-representation in official statistics, which to date still struggle to capture the extent of the female contribution to the fishing industry, especially in developing countries (Lentisco and Lee, 2015[7]). This type of widespread under-representation, very common in other ocean sectors as well, translates into a deeply unequal playing field. Traditional gender roles relegate women to activities that are less profitable and exclude them completely from resources management and decision-making processes. This exclusion undermines women's negotiation power, disregards their needs, and magnifies their vulnerabilities by negatively affecting women in terms of salary, labour force participation, economic opportunity and empowerment.

Development co-operation for sustainable ocean economies and gender equality

Development co-operation has increasingly focused on gender. The volume of ODA with a principal or significant gender component grew from USD 54 billion in 2008-09 to USD 94 billion in 2016-217, a +79% increase. In July 2019, the OECD DAC adopted the first international standard to prevent sexual exploitation, abuse and harassment in the development community, showing its strong commitment to address violence against women within the development co-operation sector itself (OECD DAC, 2019[8]).

In the ocean space, development co-operation supports gender equality with projects in several sectors including fisheries, aquaculture, maritime transport, tourism, coastal disaster risk reduction and marine protection. In each of these domains, women face specific challenges. Development co-operation needs to increasingly take into account their particular demands, rights and vulnerabilities. Data show that across ocean-based sectors, donors are increasingly mainstreaming gender equality in their policy design and implementation. In 2013-18, approximately USD 4.7 billion (26%) of the USD 17.7 billion allocated for projects to support the ocean and its economic activities integrated a gender component. As shown in Figure 4.6 the gender focus of ODA for the ocean economy varies across sectors and is highest in the fishery sector.

Figure 4.6. Share of ocean economy ODA that promotes gender equality

USD commitment (2013-18 averages)

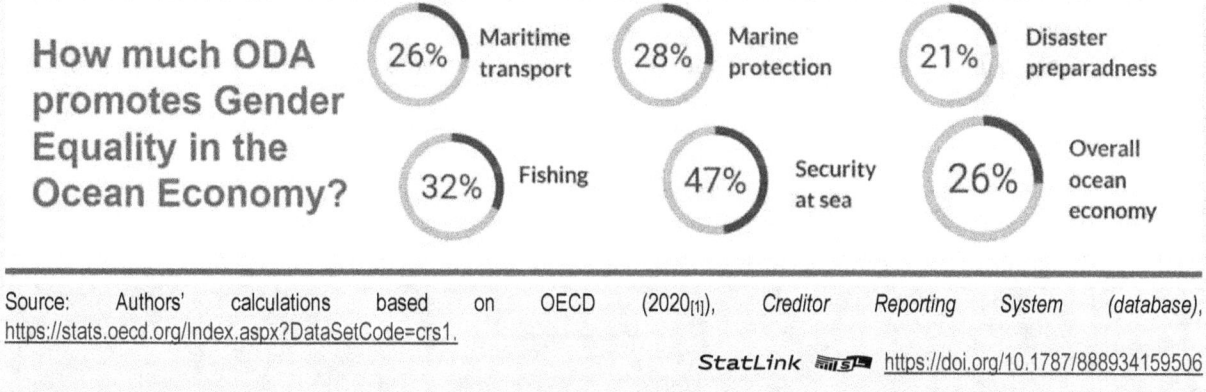

Source: Authors' calculations based on OECD (2020[1]), Creditor Reporting System (database), https://stats.oecd.org/Index.aspx?DataSetCode=crs1.

StatLink https://doi.org/10.1787/888934159506

Development co-operation is supporting women's empowerment in the ocean economy in a variety of ways, including by enhancing women's access to training, credit and production tools. Despite these positive examples, there is room for development partners to improve their support to mainstreaming gender equality and to help identify and address gender gaps in every ocean economy sector. A gender-sensitive lens needs to become a default option in ODA support to sustainable ocean economies. Development partners can further support the active participation of women in resources management and decision making, promote women's enhanced access to productive tools and resources, and advocate for an increased recognition of women's role in ocean-based sectors so that they can earn fair compensation for their work and have greater power to determine their future.

Bilateral providers account for the bulk of ocean ODA but they integrate sustainability less

In 2013-18, 60 of the 74 bilateral and multilateral development partners reporting to the CRS extended concessional finance for the ocean economy. A few partners provide the bulk of this funding. Over this six-year period, the top five accounted for 65% of the total: Japan, the European Union (EU), the International Development Association (IDA), Germany and France (Figure 4.8).

Collectively, bilateral partners committed larger volumes of concessional finance for the ocean economy than did multilateral providers, accounting for 76% of the total in 2013-18 or USD 2.2 billion on average a year. However, only 44% of bilateral ODA integrated sustainability, on average, against 72% for multilateral partners (Figure 4.7). For 43% of development partners, half or more of their ocean ODA does not integrate sustainability.

More broadly, support for the sustainable ocean economy weighs lightly on the ODA portfolios of individual providers of development co-operation, even those providing the largest volumes of ODA for the sustainable ocean economy in absolute terms. Across top providers, the share of ODA for the sustainable ocean economy ranges from 9% of the global ODA portfolio of the Global Environment Facility (GEF) to 0.2% of Germany's ODA. The shares of other providers are in between, at 2% for ADB, 1.5% for IDA and 0.9% for Japan. Across all providers, the Nordic Development Fund leads with 13.7% of its total concessional finance in 2013-18 prioritising the sustainable ocean economy.

The low prioritisation shown in the data is reflected in the widespread lack of high political commitment or specific strategies in support of the sustainable ocean economy. Fewer than two fifths of DAC members identify ocean sustainability as a priority or matter of concern in their developing co-operation strategies or policies (OECD Survey). One reason is that available data and analysis to guide policy making and interventions in this area are insufficient, both in developing countries and at development partner level. Evidence from the Sustainable Ocean Economy Country Diagnostics conducted for this report suggests that the lack of dedicated strategies and of a one-stop shop at headquarters for development co-operation on the sustainable ocean economy seems to limit development partners' ability to respond rapidly and most effectively to developing countries' requests for support in this area, even when this is a clear priority for developing countries.

Figure 4.7. Top providers of ocean economy ODA

ODA Commitments, annual averages in 2013-2018

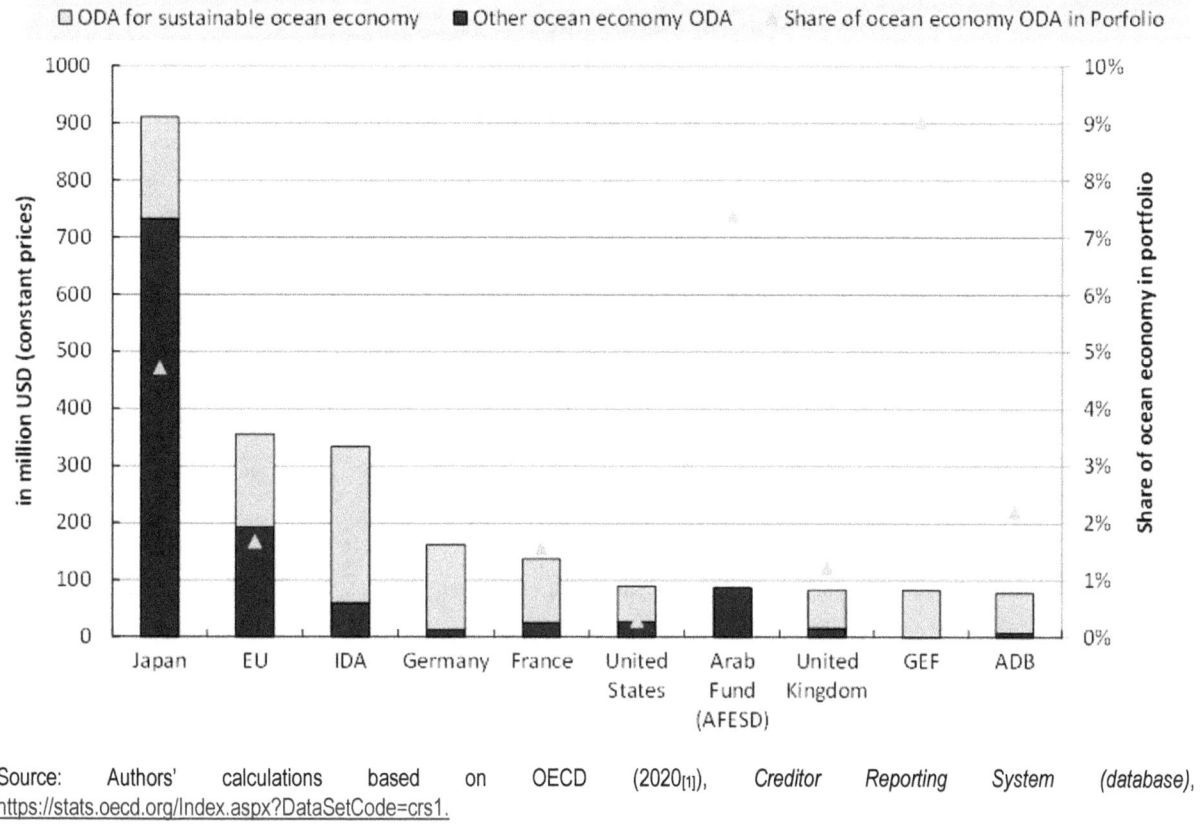

Source: Authors' calculations based on OECD (2020[1]), *Creditor Reporting System (database)*, https://stats.oecd.org/Index.aspx?DataSetCode=crs1.

StatLink https://doi.org/10.1787/888934159525

Figure 4.8. A few development partners provide the bulk of ocean economy ODA

ODA commitments, constant 2017 prices, 2013-18 cumulative shares

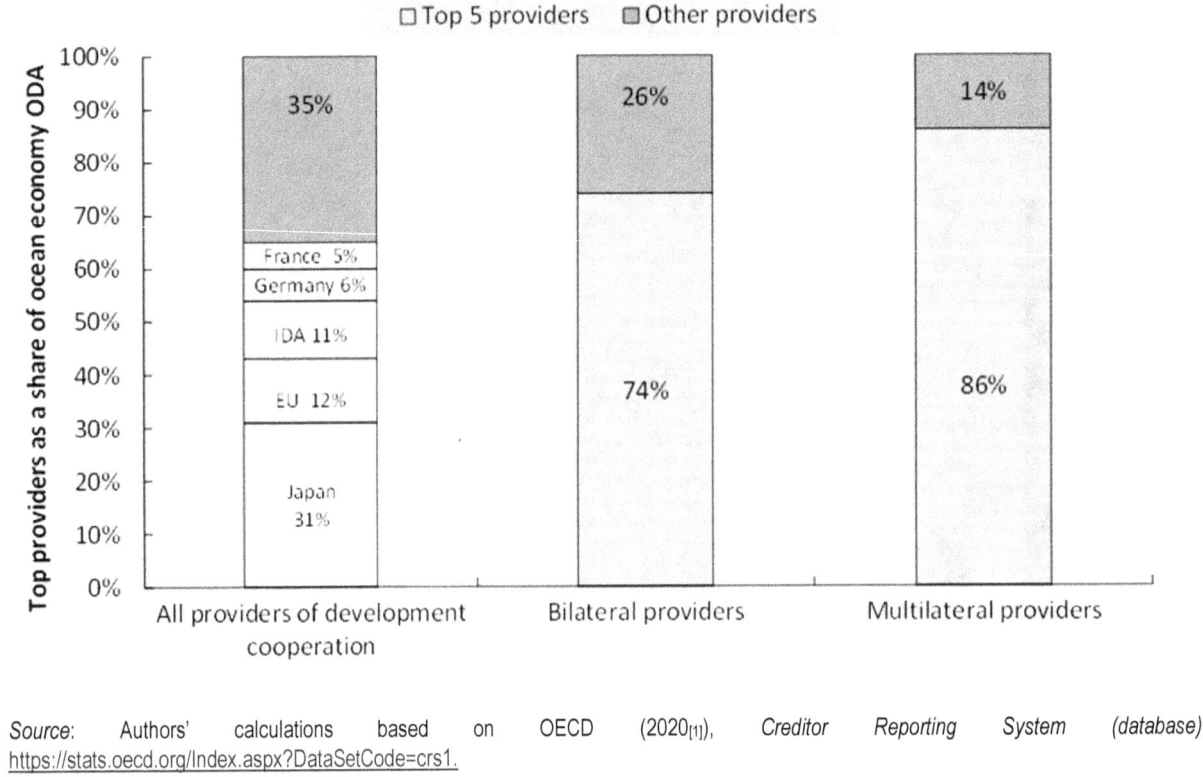

Source: Authors' calculations based on OECD (2020[1]), Creditor Reporting System (database), https://stats.oecd.org/Index.aspx?DataSetCode=crs1.

StatLink https://doi.org/10.1787/888934159544

Ocean-related ODA flows concentrates in a few recipient countries, driven by large projects that do not integrate sustainability

ODA for the ocean economy is concentrated in a few countries, with Indonesia the top recipient, followed closely by Viet Nam and Morocco (Figure 4.9).

In 2013-18, the top 20 recipients together accounted for 75% of ocean economy ODA and received, on average, allocations that were almost 10 times bigger than were received by the next 45 recipients combined (Figure 4.10). Only 4% of ocean economy ODA went to the bottom 50 recipients combined. Across country groups, over the 2013-18 period, lower middle-income countries received the largest share of ocean economy ODA (54%). Least developed countries received 31% and upper middle-income countries 15%. This breakdown resembles overall ODA allocations by country group.[3]

For many of the largest recipients of ocean economy ODA, the bulk of this assistance supports expansion of ocean-based industries without the aim to increase their sustainability. Most often these are infrastructure projects in maritime transport. Generally, only one third of the ocean economy ODA to the top recipients integrates sustainability.[4] Sustainability was incorporated to a much higher degree on average (75%) in the small allocations to the bottom 50 recipient countries (Figure 4.10). These figures indicate that ODA, to a large degree, is used dichotomously, funding either ocean-based industries with no sustainability focus in top recipients, or ocean conservation through small allocations in small recipients. These figures also suggest that increases in ocean-related ODA have not been driven by greater integration of sustainability into ocean-based industries or by larger investments towards the conservation and restoration of ocean ecosystems.

Among the top recipients, Bangladesh, India and Madagascar are exceptions in terms of the sustainability focus of ocean economy ODA; in 2013-18, the ODA received by the three countries integrated sustainability almost in their entirety (Figure 4.9). Bangladesh's high share of sustainable ocean economy ODA was primarily driven by projects promoting coastal resilience. Bangladesh, a low-lying country exhibiting high coastal vulnerability to natural hazards and high tides, received support from several donors during the six-year period to implement projects to increase the resilience of coastal population to natural disasters and climate change. In 2013, for example, IDA provided a USD 400-million concessional loan for a project of coastal embankment improvement in six coastal counties in southern Bangladesh that helped to rehabilitate polders and provide protection to an estimated 760 000 coastal inhabitants. In India, a large share of ODA for the sustainable ocean economy consisted of a USD 76-million project to establish ship recycling facilities in Gujarat. In Madagascar, the integration of biodiversity offsetting measures to protect coral reefs in the framework of the Toamasina Port Development project accounts for the relatively large share of its ODA that targets conservation and sustainable use of the ocean. The Japan International Cooperation Agency (JICA) promoted the project and provided a USD 400-million concessional loan to improve maritime connectivity in Madagascar while at the same time adopting measures to safeguard the marine ecosystems affected by this development.

Figure 4.9. Top recipients of ocean economy ODA receive small shares of funding integrating sustainability

ODA commitments, annual averages 2013-2018

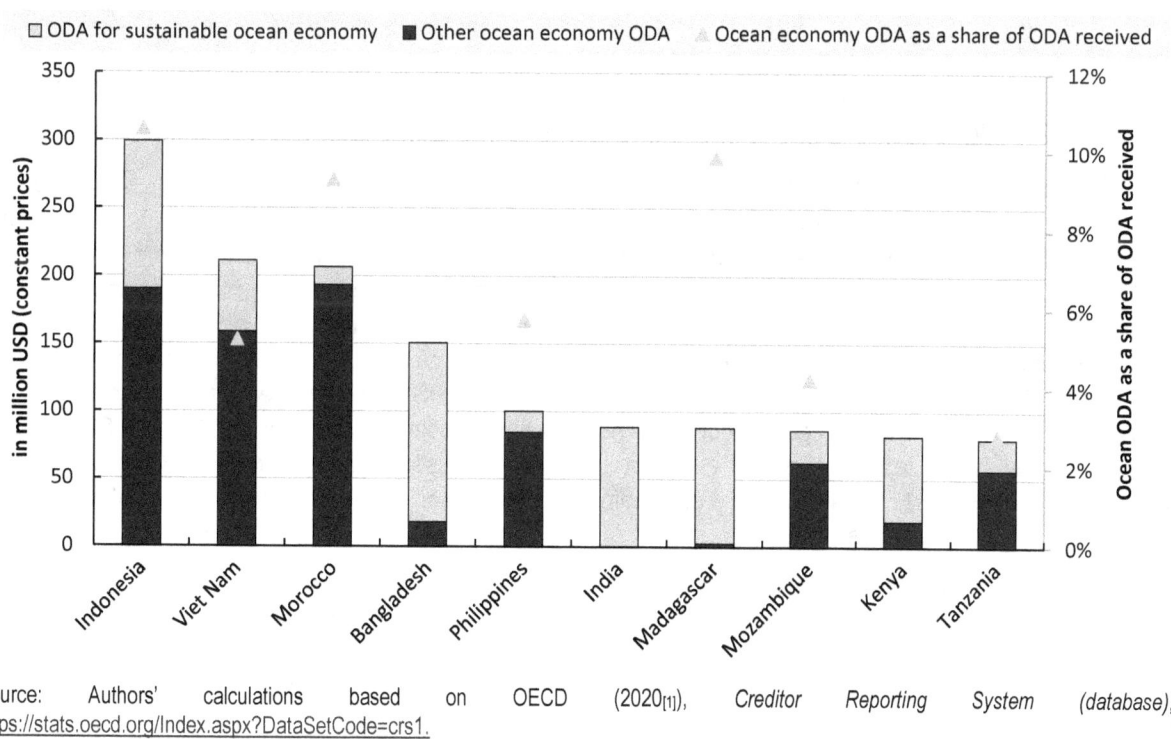

Source: Authors' calculations based on OECD (2020[1]), Creditor Reporting System (database), https://stats.oecd.org/Index.aspx?DataSetCode=crs1.

StatLink https://doi.org/10.1787/888934159563

Figure 4.10. ODA is used rather dichotomously, either funding ocean-based industries with no focus on sustainability in large recipients or ocean conservation through small allocations in small recipients

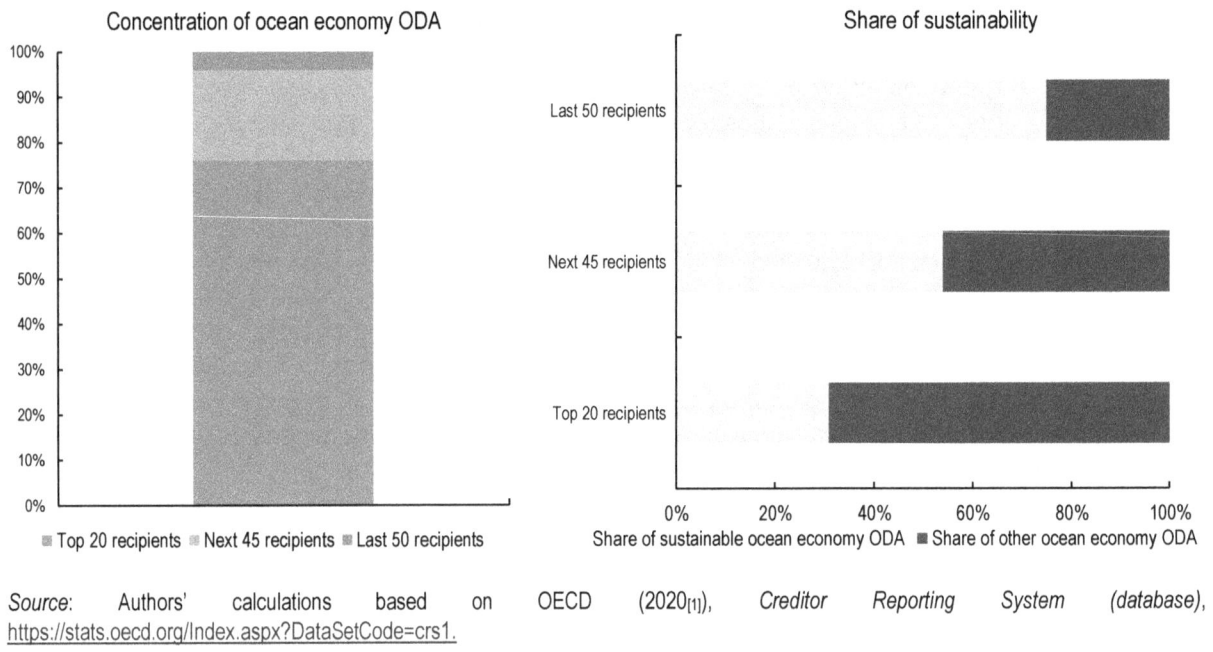

Source: Authors' calculations based on OECD (2020[1]), Creditor Reporting System (database), https://stats.oecd.org/Index.aspx?DataSetCode=crs1.

StatLink ᎙ᏚᏞ https://doi.org/10.1787/888934159582

Despite their reliance on ocean resources, small island developing states receive only a small share of ODA towards the sustainable ocean economy

Small island developing states (SIDS)' small land masses and remoteness have historically been sources of unique economic and environmental constraints hampering development (OECD, 2018[9]). But for many of these countries, sustainable ocean economies could turn their vast ocean resources – on average, more 2 000 times[5] the size of their land masses – into a driver of economic diversification, resilience and inclusive development (OECD, 2018[9]). Ocean-based sectors such as costal tourism and fisheries are already the foundation of SIDS' economic activities and livelihoods and a source of foreign exchange and employment. This strong reliance on ocean-based sectors, however, leaves SIDS particularly exposed to the increasing degradation of the marine and coastal ecosystems on which these sectors depend. To help SIDS make progress on sustainable development, support them in addressing these increasing pressures and developing new opportunities from an expanding global ocean economy, (e.g. renewable energy, aquaculture, etc.) is therefore critical.

SIDS are taking bold stances on the sustainable ocean economy and are calling on the international community to support their ambition. SIDS have identified sustainable ocean economies as an SDG accelerator, considering that investments in the sustainable ocean economy will have large multiplier effects across many other economic and social areas. These countries have become international leaders on the sustainable ocean agenda, organising regional and international events on the topic. Many SIDS also have developed blue economy strategies. The Seychelles' 2018 Blue Economy Strategic Framework and Roadmap, Mauritius' oceans economy road map, and Grenada's Costal Blue Growth Masterplan are examples. Some countries have set up dedicated institutional arrangements such as blue economy ministries (e.g. Cabo Verde, Barbados, Grenada). Some have made important commitments on the blue economy. Cook Islands declared its entire exclusive economic zone (EEZ), equivalent to 1.9 million km², a multiple-use marine protected area – the world's largest. Palau established its entire EEZ as a fully protected marine reserve, making it a no-take zone and banning all fishing and mining activities. SIDS have been pioneers in financial innovations for the ocean: the Seychelles, for instance, issued the first blue bond, as discussed in Chapter 5.

While remittances dominate SIDS' external financing flows, ODA is the largest flow of external finance for two out of five SIDS, and many are heavily dependent on ODA for the provision of basic services (OECD, 2018[9]). This reflects the fact that private investments remain limited in most SIDS, in large part because of their exposure to disaster risks, high perceived investment risks, and the often economically isolated nature and limited scalability of operations (i.e. often remote locations). These severely restrict opportunities for business development and integration in global value chains. In turn, public investments are constrained by volatile domestic revenues and limited fiscal space, since several SIDS are in debt distress or at risk of debt distress (OECD, 2018[9]).

Despite the importance of ODA in the financing landscape of SIDS, and the importance of ocean-based industries and ecosystems for their economies and livelihoods, only 5.5% of ODA to SIDS targets the ocean economy, amounting to USD 1.8 billion in 2013-18 or USD 296 million a year on average. This figure drops to 2.7% for ODA for the sustainable ocean economy in the same period, totalling USD 871 million or USD 145 million a year on average. Among SIDS, the share of ODA channelled towards the sustainable ocean economy is highest in Nauru (20%), while it accounts for less than 1% of ODA to Cuba, Cabo Verde, Haiti and Montserrat (Figure 4.11). For SIDS as a group, ODA for the sustainable ocean economy has increased since 2017 but is still below the 2014 level.

Not only is the ocean important to SIDS. SIDS are important to the ocean and the benefits that all humankind derives from it. SIDS control 30% of the world's 50 largest EEZs[6] (UN, 2014[10]). Considered custodians of the ocean, they are home to vast reserves of minerals, natural gas, fish and seafood. For instance, it is estimated that Pacific SIDS are home to the largest underwater cobalt-rich crusts, and they are already being approached by large companies for deep-seabed mining prospecting. While the science is not yet sufficiently developed to appraise the full range and scale of the impacts that such large-scale industrial operations would have on these already fragile natural ecosystems, they pose the risk of irreversible damage as well as major questions regarding appropriate governance and regulatory regimes. Development co-operation has an important role to equip SIDS with the expertise and capacities to assess and respond to these prospective projects in a way that takes into consideration both the global public good dimension and implications for the overall health of the world's ocean.

Figure 4.11. For most SIDS, ODA for the sustainable ocean economy makes for a small part of the ODA they receive

ODA commitments, annual averages in 2013-18, constant 2017 prices

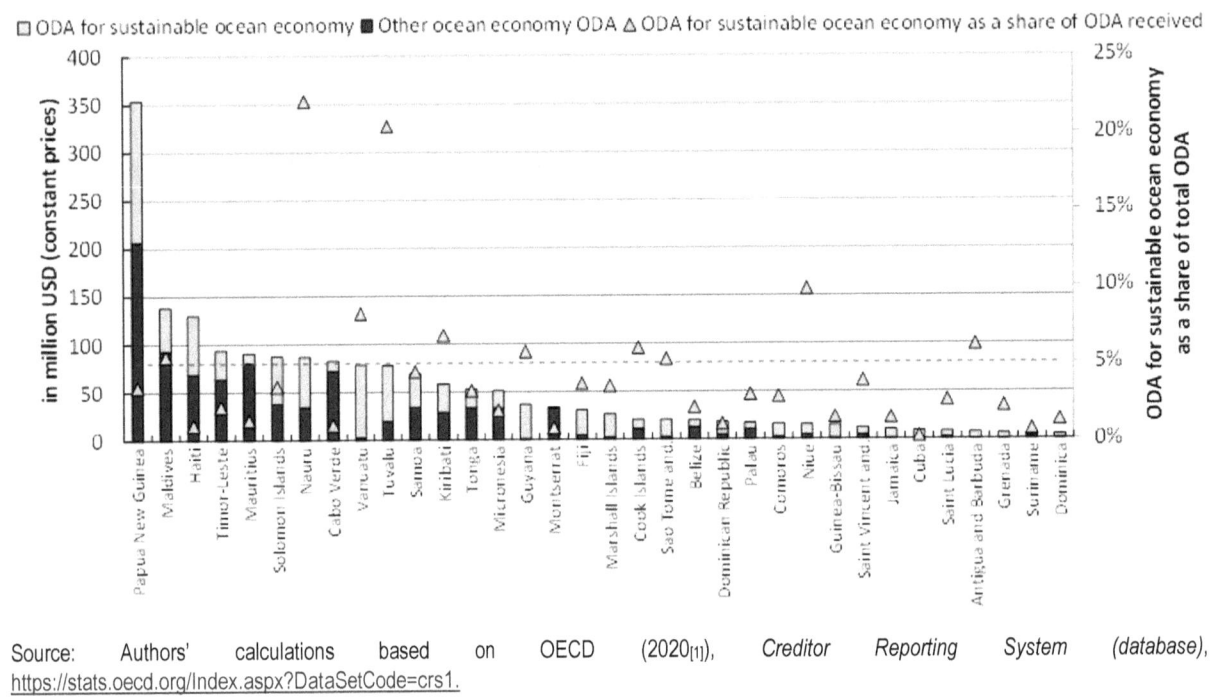

Source: Authors' calculations based on OECD (2020[11]), *Creditor Reporting System* (database), https://stats.oecd.org/Index.aspx?DataSetCode=crs1.

StatLink https://doi.org/10.1787/888934159601

Least developed countries do not prominently engage on the sustainable ocean economy

Together with SIDS, LDCs are a group of particularly vulnerable countries that the international community should make sure will be able to harness the benefits of the ocean economy. Nine of these countries also are SIDS (Annex 1A). However, coastal LDCs have been less assertive and proactive than smaller countries like SIDS in embracing the blue economy concept, largely because ocean-based sectors play a smaller role in supporting their economies and their populations' livelihoods. Nonetheless, several LDCs have established institutional arrangements and policy frameworks to use their blue natural capital more sustainably and as a driver of sustainable development. In 2017, Bangladesh created an inter-ministerial platform, the Blue Economy Cell, to develop a road map and co-ordinate initiatives among ministries on the sustainable ocean economy. Other countries, among them Cambodia and Mozambique, have folded ocean economy responsibilities into an existing ministry. In its 2013-30 National Strategic Plan on Green Growth, Cambodia, also a specific focus on blue economy development with sustainability.

The Sustainable Blue Economy Conference organised in Kenya in 2018 provided impetus to African countries, including some African LDCs, to pursue opportunities for shared prosperity from healthier oceans. At the conference, Liberia made a commitment to launch blue economy initiatives consistent with its national development plan and Pro-Poor Agenda for Prosperity and Development. To further galvanise ocean action domestically and in the region, Liberia then organised a Blue Ocean Conference in 2019 that culminated in a call for action to Liberia and the West Africa region to advance the blue economy agenda. Sudan made commitments around marine protection and the adoption of more sustainable fisheries practices.

In 2013-18, LDCs received ODA for the ocean economy in the amount of USD 4.5 billion (USD 758 million on average a year), of which 58% integrated sustainability. This represented 25% of all ocean economy ODA and 30% of the all sustainable ocean economy ODA. Over the period, ocean economy ODA to LDCs was on a downward trend until it peaked in 2017. Ocean economy ODA to LDCs predominantly targeted five sectors: maritime transport (43%), fishing (24%), coastal resilience (10%) marine protection (9%) and maritime security (3%). The five top LDC recipients together received approximately 62% of the total. More than 85% of the ocean economy ODA received by two the top five (Bangladesh and Madagascar)targets ocean conservation and sustainable use, compared to less than 40% of this ODA received by the other three (Mozambique, Tanzania and Cambodia) (Figure 4.12).

Figure 4.12. Ocean economy ODA to LDCs integrates sustainability by 58% on average

ODA commitments, annual averages in 2013-18, constant 2017 prices

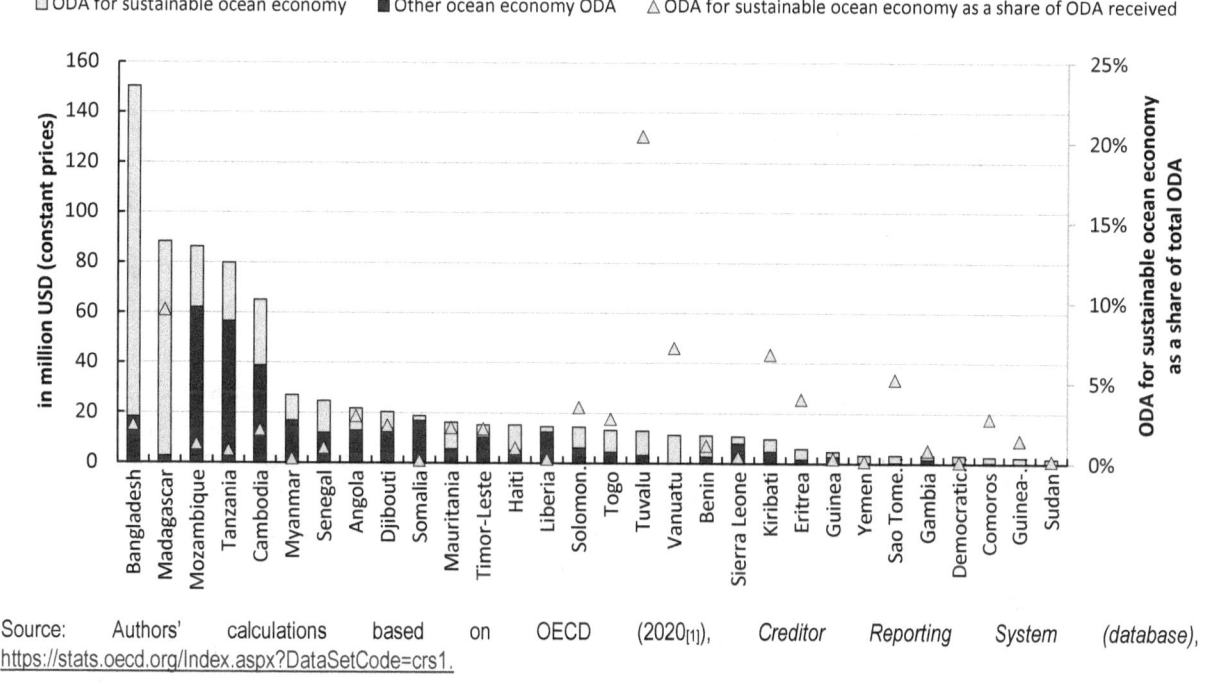

Source: Authors' calculations based on OECD (2020[1]), Creditor Reporting System (database), https://stats.oecd.org/Index.aspx?DataSetCode=crs1.

StatLink ᐊᒥᔅᒪ https://doi.org/10.1787/888934159620

Mapping the range of sustainable ODA activities for the sustainable ocean economy across key areas

Development co-operation providers have conducted several ODA projects contributing to sustainable ocean economies in developing countries. This includes ODA to support the use of new technologies to enhance sea surveillance and security; measures to increase the value of fish products through certifications of sustainability; interventions to foster sustainable tourism; and projects to reduce the environmental impact and greenhouse gas emissions from ships and port infrastructure. These examples contribute to a common understanding of what constitutes sustainable activities across ocean-based sectors and offer insights into replicable practices. They also demonstrate that, compared to defining what is "green", which is mainly associated to energy efficiency and reduced GHG emissions, sustainability in the ocean economy is a more multidimensional concept. The ocean economy is comprised of many different sectors and improving sustainability may entail a different range of actions in each one of them.

The remainder of this chapter examines ODA allocations across the sectors of the sustainable ocean economy and then provides a more granular analysis of the range of sustainable projects in six specific areas/sectors of the sustainable ocean economy. The economic trends and domestic policy instruments to increase sustainability in most of these areas/sectors are explored in Chapters 2 and 3. Here, the range of ODA projects conducted to enhance their sustainability is spelled out to contribute to a common understanding of what constitutes sustainable interventions in each of them and provide examples of replicable practices. The six areas span existing and emerging ocean-based sectors and include an area specifically focused on ocean conservation and restoration. They were chosen for their general relevance to developing countries, although it is acknowledged that each country will prioritise a different set of ocean-based sectors according to its assessment of opportunities, comparative advantage and national interests. Therefore, the six areas by no means constitute a blanket prescription and were selected to provide more detailed information. The chapter also suggests how efforts to enhance sustainability might be stepped up across these six areas.

Allocations across sectors: A narrow approach?

ODA for the sustainable ocean economy is concentrated in three sectors: maritime transport and infrastructure (22.1% of the total in 2013-18), fisheries (21.5%), and marine protection (20.9%). ODA for the ocean economy is even more concentrated in maritime transport, which accounts for 47% of the total, and with 17% going to fisheries and 11% to marine protection. Support to these sectors integrates sustainability to varying degrees. Only 24% of ODA for maritime transport and infrastructure integrates sustainability compared to 66% for fisheries, although these shares are highly variable from one year to the next.

This may be too narrow an approach of development co-operation to the sustainable ocean economy, considering that realising the potential of the sustainable ocean economy will require investing in an array of sectors to both expand the sustainability of existing ocean-based sectors and harness opportunities from emerging sectors. The potential of each of these industries varies across developing countries (Chapter 2). Providers of development co-operation will need to prioritise their support accordingly, in line with developing countries' assessment and prioritisation of opportunities, comparative advantage and national interests.

Some DAC members, both at home and abroad, have embraced marine spatial planning (MSP) as a tool for an integrated management of coastal and marine resources to balance multiple objectives in the conservation and sustainable use of these resources (Chapter 3). However, development co-operation support to MSP remains limited; only 17 of the more than 600 conservation projects conducted in 2013-17 focused on it either in part or in full. Yet MSP presents several challenges that the international community could help address, including through the development and use of appropriate data, models and decision support tools to inform the planning process.

Figure 4.13. How development co-operation is helping enhance sustainability across six sustainable ocean economy areas

Note: USD amounts are annual averages in the 2013-18 period
Source: Authors

Conserving and restoring the ocean

Alongside the mainstreaming of a more sustainable use of ocean resources across ocean-based industries, specific actions are required to conserve and restore marine ecosystems. As noted, the marine protection sector receives the third largest amount of ODA for the sustainable ocean economy, after maritime transport and fisheries. In 2013-18, USD 312 million a year on average was allocated for marine protection, representing 6.4% of total ODA for general environmental protection in this period.

Interventions in this area of the sustainable ocean economy ranged from support to develop comprehensive, cross-sectoral approaches to the conservation and sustainable use of biodiversity and building capacity for local personnel to implement marine conservation monitoring and management strategies to specific conservation and restoration projects. In 2013-18, 14% of all ODA for marine conservation funded the establishment and management of marine protected areas (MPAs). In addition, 67% of marine protection projects integrated resilience, mainly through nature-based solutions such as mangroves.

One example of a cross-sectoral approach is the project for rehabilitation and sustainable management of mangroves, developed in 2007-10 as a partnership between Indonesia and JICA. The project combines mangroves restoration with eco-tourism and brings co-benefits to fishing communities. Mangroves were planted in 253 hectares across different parts of Bali and Lombok islands, where a mangrove forest had been cut to create space for the construction of fish ponds. Among other outputs, the rehabilitated mangrove forest provides fishery resources such as mangrove crabs and fishes. The project also established the Mangrove Information Center, which promotes eco-tourism and environment education.

In addition to cross-sector approaches, cross-country approaches may be needed due to the transboundary nature of natural marine assets. A positive example is the Coral Triangle Initiative, a partnership between the governments of Indonesia, Malaysia, Papua New Guinea, Philippines, Solomon Islands and Timor-Leste with support from GEF, Australia, the United States, ADB and other development partners. The Coral Triangle is one of the greatest centres of biodiversity on Earth, containing more than 75% of the known coral species and home to 363 million people, of whom 141 million live within 30 kilometres of a coral reef (Indonesia Ministry of Marine Affairs and Fisheries, 2019[11]). The work of the Initiative is organised around the themes of assessment and action for threatened species, climate change adaptation, ecosystems-based management of fisheries, and MPAs.

Besides supporting specific marine conservation and restoration projects, development co-operation can support the enabling conditions for attracting more private contributions for marine conservation and restoration, for example from the tourism and fisheries sectors (Spergel and Moye, 2004[12]). For-profit investments in conservation have recently been discussed as a potential solution for global funding gaps in conservation finance (Huwyler et al., 2014[13]) and in this context, ODA could play a role to "subsidise demonstration projects, reduce financial risk through collaterals, and promote low profit but high co-benefit projects" (Vanderklift et al., 2019[14]). However, generating profitable returns for investors from conservation projects is challenging, and private investments in conservation remain small, marginal and geographically constrained (Dempsey and Chiu Suarez, 2016[15]).

Sustainable and traceable seafood

Several providers of development co-operation have a long history of providing support to the fisheries sector in developing countries, not least because it contributes so substantially to livelihoods and food security. Among ocean economy sectors, fisheries is the second largest recipient of ocean economy ODA in 2013-18, receiving 21.5% of the total, or an annual average of USD 488 million. Of this amount, 12%, or USD 56.6 million on average a year, was allocated for aquaculture projects in recognition that aquaculture rather than increased wild catch will drive most fish production growth in the future (Chapter 2) (FAO, 2018[16]).

ODA to the fisheries sector incorporates sustainability concerns to a greater extent (66%, or USD 321 million on average a year in 2013-18) than ODA to other ocean-based sectors. This reflects development partners' longstanding awareness of the need to balance conservation and sustainable use of fisheries resources. Development co-operation interventions for fisheries began moving away from an exclusive focus on production growth already in the 1990s to incorporate conservation, resource management and sustainable fisheries practices. This shift was driven by a growing awareness around fish stock depletion and the increase in illegal, unreported and unregulated (IUU) fishing, a major factor behind overfishing (Carneiro et al., 2019[17]; Widjaja, Long and Wirajuda, 2020[18]).

Development co-operation interventions for improved fisheries management show that co-benefits from conservation and sustainable use of resources can effectively be achieved and that improved fisheries management is associated with higher catches and positive economic outcomes. For instance, the United States Agency for International Development (USAID) programme in the Philippines, Ecosystems Improved for Sustainable Fisheries, or ECOFISH, resulted in a 24% increase in fisheries biomass and a 12% increase in employment through the integration of marine conservation and fisheries management (OECD Survey).

Several development projects enhanced value addition of fish products. One of these, Sustainable Market Access through Responsible Trading of Fish in Indonesia, also called SMART-Fish Indonesia, is a partnership between Indonesia, the United Nations (UN) Industrial Development Organisation and the Swiss Agency for Development and Cooperation to enhance the value of fish products through improved fisheries management practices, eco-labelling and new technologies for product traceability.

Effective fisheries management requires improved sea surveillance and monitoring as well as the availability of science and data on fisheries. The development community is providing support to both areas, although evidence from the Sustainable Ocean Economy Country Diagnostics conducted for this report suggests that country needs have been only partially met and remain significant. An example of co-operation to increase access to fisheries data include Italy's support to Palau and other small islands for assessment of potential effects of climate change on the distribution, long-term movements and local fisheries productivity of pelagic and near-shore resources. Another example is the training provided by Korea provided in Asia and Africa for ocean observation and hydrographic surveying to strengthen the evidence base on ocean observation, hydrographic surveying technology and climate change. Efforts to improve sea surveillance and monitoring in response to widespread piracy and illegal activity at sea have also included training for coast guards and maritime authorities and support to implement vessel monitoring systems and space technologies for mapping ocean resources.

Development partners are also supporting the use of innovative technologies such as blockchain to increase traceability of fish, help combat IUU fishing and curb human rights violations, including human trafficking and slavery on fishing boats. An example is a pilot project in the Pacific islands tuna industry launched by the World Wildlife Fund in Australia, Fiji and New Zealand; ConsenSys, a technology innovator based in the United States; the technology implementer, TraSeable; and Sea Quest (Fiji) Limited, a tuna fishing and processing company. The partnership will use blockchain technology to track tuna from bait to plate. USAID has partnered with the Walton Family, Packard and Moore Foundations on the Seafood Alliance for Legality and Traceability, or SALT, a global alliance that brings together industry, governments, traceability technology companies and civil society to accelerate learning and support collaboration around traceability approaches to legal and sustainable seafood.

Fisheries development co-operation interventions, however, need to better assess and address equality and distributional effects. Challenges remain to the effective involvement of coastal communities in fisheries management schemes and in the development of alternative income sources for fishers to effectively reduce overfishing. Greater effort also is needed to achieve gender equality in fisheries management schemes. As noted in Box 4.2, only 32% of development co-operation interventions in the fisheries sector in 2013-18 addressed gender equality.

Finally, fisheries subsidies can have impacts on the sustainability of fisheries, including outside the boundaries of the subsidising country's waters. As SDG Target 14.6 underscores, and as discussed in ongoing negotiations in the World Trade Organization, it is important to ensure that support policies do not encourage overfishing, illegal, unregulated and unreported (IUU) fishing, or other fishing practices that destroy ocean ecosystems and compromise the sustainability of resources. Policy coherence and commitment from the development co-operation community are key to ensuring donors' national or regional fisheries policies either support or at least do not undercut their development policies. Pro-active phasing out of potentially harmful fisheries subsidies should be encouraged, without necessarily waiting for an international agreement. Policy coherence is recognised as an important part of the Sustainable Development Agenda and SDG Target 17.14 calls on all countries to "enhance policy coherence for sustainable development" to strengthen the means of implementation.

High seas and straddling stocks are overfished at twice the rate of those within national jurisdictions (FAO, 2014[160]). While developing countries lack the resources (e.g. large vessels) and technologies necessary to access deep-sea fisheries (Widjaja, Long and Wirajuda, 2020[18]), they are nonetheless affected by high-sea fishing, as ocean currents connect what happens on the high seas to what happens close to shore. The UN General Assembly is currently negotiating a new international treaty for the conservation and sustainable use of biodiversity beyond national jurisdiction (BBNJ). The new treaty could provide an excellent opportunity to curb and address the impacts of overfishing on the high seas and enable a more inclusive and sustainable use of high seas resources to the benefit of all countries.

Despite the growing importance of aquaculture, the share of fishery ODA targeting aquaculture activities is small, partly because the sector is largely financed via large private investments, and declined in 2013-18. Despite limited ODA flows towards aquaculture, positive examples of development co-operation in this area exist. Norway's Fish for Development programme provides vocational training to the local labour force in developing countries to enable the development of large-scale aquaculture plants and supports research on fish health to enable the development of new aquaculture species. PROBLUE is also focusing part of its financing on sustainable aquaculture in its first annual work plan and budget. Out of the total budget of USD 8 million allocated for improved fisheries governance, approximately USD 1.4 million targets sustainable aquaculture with a specific focus on producing analytical work and building local capacity for disease prevention and biosecurity.

Sustainable tourism

Despite the importance of the tourism sector for many developing countries and the need to enhance the sustainability of the industry, as discussed in Chapters 2 and 3, only a tiny part of ODA is allocated to this sector. The small volume of ODA to this sector is partly connected to the heavy reliance of the sector on large private investments. In 2013-18, ODA for coastal tourism amounted to USD 9.7 million on average a year, of which 64% integrated sustainability. Development co-operation approaches to enhance sustainability of the tourism sector currently range from support for national strategies of sustainable tourism to ecosystem payment schemes associated to MPAs and support for certifications or eco-tourism projects.

Development co-operation could provide developing countries broader support for tourism policies and approaches that foster forms of tourism that reduce environmental degradation and favour strong local returns for inclusive, sustainable development. These are critical as mass tourism – the prevailing form of tourism – is associated with significant environmental degradation and large financial leakages in many developing countries (Chapters 2 and 3).

To enhance the social sustainability of the tourism sector, providers of development co-operation can support measures to foster backward and forward linkages with the rest of the economy and promote greater local ownership. Providers can also encourage regional co-operation, for instance on fees and taxes on large cruise companies for docking and landing, to avoid a race to the bottom. Through education, training and support for the introduction of specific requirements, development partners can support local tour operators, locally owned business, and local suppliers in other sectors such as agriculture, food processing, handicrafts, trade, transport, and recreation and entertainment. Development co-operation providers can also support public and private investments in low-carbon transport options and the construction of resource-efficient tourism infrastructure (OECD, 2018[19]).

Finally, providers of development co-operation could further support developing countries in testing innovative ways to raise tourist awareness and to channel some tourism revenues towards the conservation of natural assets. To raise awareness, Palau changed its immigration laws in 2017 and now requires that visitors sign a pledge upon entry, which is stamped in their passports, to act in an ecologically responsible way and protect Palau's environment and culture for the next generation. This compulsory promise is made directly to the children of Palau to preserve the country that is their home. Funding in support of the initiative came from traditional donors such as Italy as well as non-governmental organisations, private donors, local businesses and even community fundraisers, and the government of Palau. In Mexico in 2018, resources from the private sector were used to create a Coastal Zone Management Trust through a partnership between the government of Mexico, the UN and other development partners Taxes on the tourism industry finance the Trust and are used to maintain and conserve 60 kilometres of reef and beaches in the area of Cancun and Puerto Morelos. The Trust proceeds also cover parametric insurance for coral reefs and beaches to protect the natural capital, businesses and communities in the event of natural disasters (Chapter 5).

Curbing ocean pollution from land

As highlighted in Chapters 1 and 3, pollution from land-based activities such as agricultural runoffs and waste is a key pressure on marine ecosystems. In 2013-18, development partners provided a total of USD 11 billion (USD 1.8 billion on average a year) for land-based interventions that reduce negative impacts on the ocean. Most of this funding (89%) targeted the improvement of waste management systems, while only 1% addressed agricultural runoffs. Concerns about the adverse impacts of marine plastic litter have led to a growing number of high-profile initiatives. Approximately USD 278 million (USD 46 million on average a year) of ODA was allocated to address marine plastic pollution in 2013-18. The annual average is expected to grow, as many commitments were made in 2019 and 2020 that are not yet reflected in the most recent, 2018, ODA figures.

As the G20 Action Plan on Marine Litter underscores, the leakage of plastics litter into the ocean can be addressed in a variety of ways throughout the plastics lifecycle. Among these are source reduction through the use of alternative materials, enhanced waste collection and recycling, and clean-up and remediation activities such as beach clean-ups and technology to collect plastics from the ocean. Effectively addressing marine plastics litter will require a combination of these approaches (Chapter 3). Development co-operation can provide support at all the various stages. At present, development co-operation to help curb ocean pollution is mainly focused on supporting central and local governments to enhance waste management systems and supporting local communities to improve collection and recycling through the production of handicrafts and construction materials from plastic waste.

For example, WasteAid, an UK Aid-funded initiative, works with communities in low-income countries to raise awareness on the problems caused by plastic pollution and develop practical alternatives to open dumping and burning of plastic waste, thus helping convert plastic waste into a source of sustainable business. The GEF Small Grants Programme, implemented by the UN Development Programme, has worked with local communities to test innovative solutions for plastic waste management through a circular economy approach, which promotes closed-loop production and consumption. The programme specifically targets the poorest and most vulnerable people in the community – women, youth and disabled people, among others – and works with them to develop circular economy approaches that combine traditional local knowledge, the application of modern science and technology, awareness raising, and advocacy. A project in India offers a positive example of this Nearly 700 ragpickers, including many socially marginalised and illiterate women, were trained in waste collection and recycling activities. As a result, approximately 10 tonnes of plastic waste are collected at five recovery centres in Bhopal every day, and cement industries in and around the city recycle the plastics, thus contributing to provide a source of livelihood.

The Australian government, in a good example of international policy coherence for a sustainable ocean economy, is implementing a phased waste export ban to ensure that Australia's waste is managed appropriately and will not create environmental problems in recipient nations. Global agreements and standards for waste management could be a useful addition to policy levers to reduce pressure on the ocean.

Greening ports, shipping and maritime transport

Ocean economy ODA primarily targets the maritime transport and shipping sector. In 2013-18, the sector received 47% of the total ocean economy ODA, or USD 1.4 billion a year on average. Most ODA to the sector comprised loans for the development and rehabilitation of port infrastructure. However, only 24% of the ODA to this sector in 2013-18 (USD 329 million on average a year) integrated sustainability. In other words, not even a quarter of this funding contributes to sustainable ocean economies.

ODA funded a range of activities to enhance the sustainability of maritime transport and shipping, largely in relation to infrastructure: from mitigating and offsetting the negative environmental impacts of maritime infrastructure projects and building climate-resilient port infrastructures to upgrading ship recycling facilities, and improving waste management in ports and harbours. Some projects focused on enhancing developing countries' domestic capacities to implement International Maritime Organization recommendations through education and training as well as through technical assistance to support countries in assessing emissions from their domestic fleet and testing low-carbon technologies for domestic shipping.

These projects demonstrate that the development community can, and has already successfully started to support developing countries in making progress along several of the dimensions that constitute enhanced sustainability of the maritime transport and shipping sector. However, the large share of investments targeting this sector and the relatively small share of these investments that integrates sustainability indicate that the development community needs to focus more attention on sustainability. Particularly neglected areas seem to be around integration of climate-resilience in port infrastructure, including the assessment of sea-level rise and natural disasters in the context of changing ocean and climate conditions. This was found to be the case in most of the Country Diagnostics conducted for this report.

Ocean economy ODA figures do not adequately address the social sustainability of ODA interventions in this sector, and further work will be needed in this area in the future. Broadly speaking, social sustainability of the maritime transport and shipping sector can be thought of as relating to the capacity of port and maritime transport investments to generate shared benefits for the local population and thus drive inclusive development and to the capacity of such investments to enhance gender equality in the sector. An example that emerged from the Country Diagnostics conducted for this report is the effort by Indonesia to enhance inter-island connectivity. By subsidising goods shipment lines – essentially sea toll routes – the country helped to hold down basic commodities prices in disadvantaged outer islands, with a positive contribution to social inclusion. With regard to gender equality, only 26% of the ODA to the maritime transport and shipping sector in 2013-18 promoted gender equality.

Harnessing renewable ocean energy

In net importer countries, as discussed in Chapter 2, the cost of fossil fuels is a heavy burden on government budgets, makes business and living expensive, and disproportionally affects poorer segments of the population (World Bank, 2019[20]). Development co-operation can provide support to assess the potential of marine renewable energies and help bear the upfront costs of investments in marine renewable energies.

In 2013-18, development partners allocated only USD 29 million on average a year for marine renewable energy in developing countries. But there have been several positive examples. One is the support from Italy to the Ocean Energy Resources Assessment of the Maldives, which is meant to appraise the energy potential of the sea currents of the Maldivian archipelago with the aim of identifying technological solutions for their exploitation. In another example, the EU is supporting the Cabo Verde 2018-40 Masterplan for Energy, under which optimal energy mixes are identified for each island. The master plan assessed that offshore wind energy production is not feasible due to strong currents and exposure to hurricanes, but determined that sun, wind and ocean waves – all found in abundance in Cabo Verde – are potentially exploitable energy sources that would lead to lower energy costs in the long run. Development c-operation providers are currently working to address the high perceived investment risks and relatively low financial viability related to the relatively small size of the Cabo Verde market.

Development co-operation interventions are also establishing synergies across ocean-based sectors, for instance through the use of renewable energy sources such as solar energy for fisheries and water desalinisation projects. In the Pacific, Italy supported Kiribati through the project, Solar Off-Grid Systems for Outer Islands fish centres. The first phase of the project has provided local fishing communities continuous access to chillers to use for their catches, resulting in increased profits and the substitution of fossil fuel with renewable energy sources. With the second phase, ten more fishing centres in the remote islands will be electrified with solar energy. In the Caribbean, Italy is supporting the construction and installation of a photovoltaic-powered salt water reverse osmosis desalination plant in Grenada The plant, to be built in the rural area of the island of Carriacou, purify, collect and distribute about 300 m3 of water per day, alleviating the shortage of fresh water on the island through the use of renewable energy.

Enhancing the impact of development co-operation for sustainable ocean economies

As ecosystems change, fish stocks move and trade routes open or shift, the state of the ocean is not static, and ocean science and knowledge of it are constantly evolving. Therefore, it is essential that policy, decision making and resource management approaches, including those related to development assistance, reflect the most recent science and facts-based evidence.

To effectively support developing countries as they face new opportunities and challenges from the ocean economy, development co-operation needs to strengthen its evidence-base and tailor its toolkits and approaches so that it can provide coherent support and for sustainable ocean economies and maximise its impact. Specific priorities include the following:

- Support developing countries to develop a coherent, unified vision and direction for the sustainable ocean economy, where the complexity of inter-sectoral interactions is understood, environmental, social and economic values are integrated, and adequate resources are mobilised across sectors.

- Prioritise support to create decent local jobs, protect livelihoods, conserve nature and promote community-based approaches for the management of ocean resources to effectively help make both emerging and existing ocean-based sectors catalysts for long-term, inclusive and sustainable development. In established sectors, development co-operation should focus support on correcting the trends of financial leakages, economic exclusion and environmental degradation. In emerging sectors, support can focus on helping countries assess and balance the risks and rewards of new economic opportunities to effectively integrate, from the outset, community interests and environmental concerns in decision-making and achieve a sustainable use of resources.

- Track ODA for the sustainable ocean economy and its impacts based on common definitions and clear principles. The tracking of ocean-relevant ODA should become an integral part of the regular monitoring of ODA flows in order to provide transparency and accountability of ODA flows and promote mutual learning on the most effective ODA interventions and approaches for supporting sustainable ocean economies. An official taxonomy of ODA for the sustainable ocean economy could be developed to guide the tracking and monitoring of ODA flows and that include gender equality and social sustainability criteria.
- Explore new development co-operation schemes fit for transitioning to sustainable ocean economies, such as:
  - New co-operation schemes that take into account the global public good nature of ocean resources. The exploitation of marine and seabed resources for new ocean-based extractive sectors can produce short-term revenues for individual developing countries. However, financial gains can be highly concentrated and difficult to reconcile with inclusive development. Destructive environmental impacts, meanwhile, could be huge and extend well beyond national borders, with global consequences for the ocean's ability to regulate climate, store carbon, and provide livelihoods and food. New international development schemes might be needed to compensate developing countries for foregone revenues and ensure an international cost-sharing mechanism for the protection of the ocean as a global public good. For instance, a REDD+ – Reducing Emissions from Deforestation and Forest Degradation – scheme for the ocean could be explored.
  - Development co-operation schemes to strengthen developing countries' expertise and engagement capacity on new opportunities and counterparts on the ocean economy. An ocean for development co-operation scheme could help developing countries more effectively manage their commercially exploitable marine resources by providing support for achieving fair commercial deals and concessions. Such support should also focus on assisting developing countries assess the risks and potential gains from new market opportunities, including by involving coastal communities. These schemes would be very relevant at a time when the value of ocean resources is increasing as they open up for commercial exploitation, especially through emerging industries such as marine biotechnology and pharmaceuticals.

Beyond development co-operation policies, a wide range of policies affect the sustainability of ocean economies, from fisheries agreements and green transition targets to patent regimes for pharmaceuticals. Fostering further analytical work and policy dialogue on selected aspects of international policy coherence for a sustainable global ocean economy could ensure that its benefits are harnessed for all countries. Specific priorities include the following:

- Strengthen the independent study on how the sustainability of ocean economies in developing countries is affected by policies beyond development co-operation, such as policies on fisheries, tourism, investment and finance.
- Foster evidence-based dialogue across countries on the impact of policies beyond development co-operation on the sustainability of developing countries' ocean economies.

References

Dempsey, J. and D. Chiu Suarez (2016), "Arrested development? The promises and paradoxes of 'selling nature to save it'", *Annals of the American Association of Geographers*, Vol. 106/3, pp. 653-671, http://dx.doi.org/10.1080/24694452.2016.1140018. [15]

Expert Group for Aid Studies, S. (ed.) (2019), *Fishing Aid: Mapping and Synthesising Evidence in Support of SDG 14 Fisheries Targets*, https://eba.se/en/rapporter/fishing-aid-mapping-and-synthesising-evidence-in-support-of-sdg-14-fisheries-targets/11447/. [17]

FAO (2018), *The State of World Fisheries and Aquaculture 2018*, Food and Agriculture Organization (FAO), Rome, http://www.fao.org/3/I9540EN/i9540en.pdf. [16]

Griggs, D., M. Nilsson and A. Stevance (eds.) (2017), *A Guide to SDG Interactions: From Science to Implementation*, http://dx.doi.org/10.24948/2017.01. [5]

Huwyler, F. et al. (2014), *Conservation Finance: Moving beyond Donor Funding Toward an Investor-driven Approach*, Credit Suisse, Zurich, https://www.cbd.int/financial/privatesector/g-private-wwf.pdf. [13]

Indonesia Ministry of Marine Affairs and Fisheries (ed.) (2019), *Assessment of Threatened Species in the Coral Triangle Region: Indonesia*, Coral Triangle Initiative, Manado, Indonesia, http://www.coraltriangleinitiative.org/sites/default/files/resources/Indonesia_Assessment%20of%20Threatened%20Species.pdf. [11]

Lentisco, A. and R. Lee (2015), *A Review of Women's Access to Fish in Small-scale Fisheries*, Food and Agriculture Organization, Rome, http://www.fao.org/3/a-i4884e.pdf. [7]

OECD (2020), *Creditor Reporting System (database)*, https://stats.oecd.org/Index.aspx?DataSetCode=crs1. [1]

OECD (2018), *Making Development Co-operation Work for Small Island Developing States*, OECD Publishing, Paris, https://dx.doi.org/10.1787/9789264287648-en. [9]

OECD (2018), *OECD Tourism Trends and Policies 2018*, OECD Publishing, Paris, https://dx.doi.org/10.1787/tour-2018-en. [19]

OECD (2016), *The Ocean Economy in 2030*, OECD Publishing, Paris, https://dx.doi.org/10.1787/9789264251724-en. [3]

OECD DAC (2019), *DAC Recommendation on Ending Sexual Exploitation, Abuse, and Harassment in Development Co-operation and Humanitarian Assistance: Key Pillars of Prevention and Response*, OECD Publishing, Paris, https://www.oecd.org/officialdocuments/publicdisplaydocumentpdf/?cote=DCD/DAC(2019)31/FINAL&docLanguage=En. [8]

Siles, J. et al. (2019), *Advancing Gender in the Environment: Gender in Fisheries - A Sea of Opportunities*, International Union for Conservation of Nature, Gland, Switzerland, https://portals.iucn.org/union/sites/union/files/doc/2019-iucn-usaid-fisheries-web.pdf. [6]

Spergel, B. and M. Moye (2004), *Financing Marine Conservation: A Menu of Options*, World Wildlife Fund, Washington, DC, http://awsassets.panda.org/downloads/fmcnewfinal.pdf. [12]

Stafford, R. and P. Jones (2019), "Viewpoint- Ocean Plastic Pollution: A convenient but distracting truth?", *Marine Policy* 103, pp. 187-191, https://doi.org/10.1016/j.marpol.2019.02.003. [2]

UN (2015), *Transforming Our World: The 2030 Agenda for Sustainable Development*, United Nations General Assembly, New York, https://sustainabledevelopment.un.org/post2015/transformingourworld/publication. [4]

UN (2014), *International Year of Small Islands Developing States 2014 (webpage)*, https://www.un.org/en/events/islands2014/didyouknow.shtml. [10]

Vanderklift, A. et al. (2019), "Constraints and opportunities for market-based finance for the restoration and protection of blue carbon ecosystems", *Marine Policy* 107, https://doi.org/10.1016/j.marpol.2019.02.001. [14]

Widjaja, S., T. Long and H. Wirajuda (2020), *Illegal, Unreported and Unregulated Fishing*, World Resources Institute, Washington, DC, http://www.oceanpanel.org/iuu-fishing-and-associated-drivers. [18]

World Bank (2019), *Going Global: Expanding Offshore Wind to Emerging Markets*, http://documents.worldbank.org/curated/en/716891572457609829/Going-Global-Expanding-Offshore-Wind-To-Emerging-Markets. [20]

Annex 4.A. Methodology for estimating ocean-relevant ODA

Currently, there is no internationally agreed definition of the sustainable ocean economy. Several different terms are often used interchangeably, but without a clear understanding of how they differ and where they overlap. In consequence, no predefined classification is available for identifying which official development assistance (ODA) activities can be considered as either directly contributing or indirectly relevant to achieving a sustainable ocean economy.

This report takes a practical approach to tracking ODA spending and activities that are relevant to the ocean. This approach is articulated around three ocean-relevant ODA indicators.

Annex Figure 4.A.1. Procedure to estimate the three ocean-relevant indicators

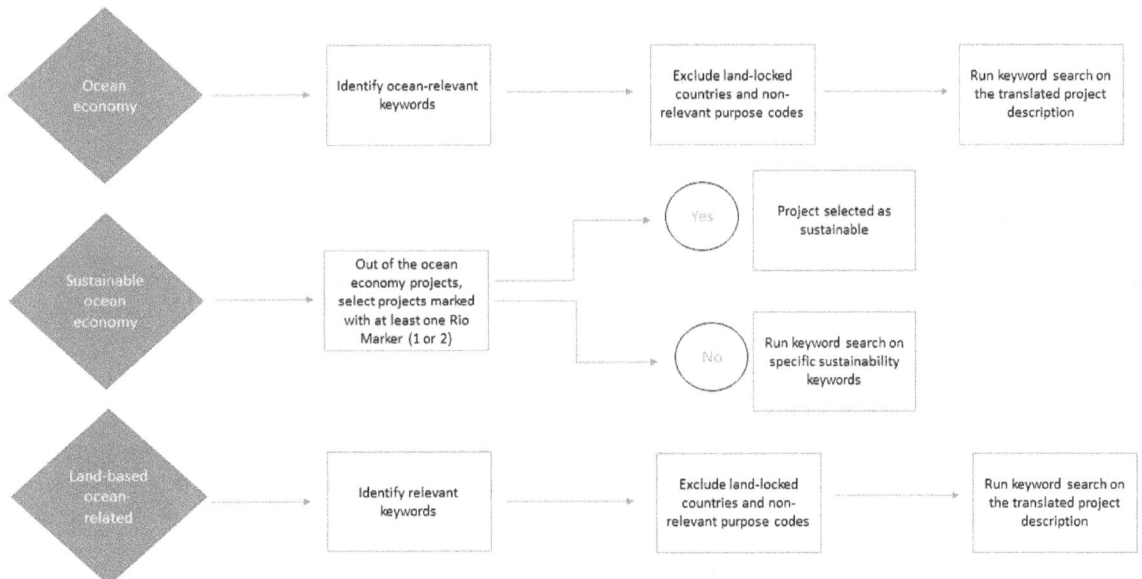

Source: Authors

ODA in support of the ocean economy irrespective of whether the support explicitly takes sustainability considerations into account

Identification of these indicators draws on the OECD definition of the ocean economy in the 2016 report, (*The Ocean Economy in 2030*, as "the sum of the economic activities of ocean-based industries, and the assets, goods and services of marine ecosystems" (OECD, 2016[3]). Annex Figure 4.A.2 presents this definition as a conceptual framework and illustrates the two components captured in Indicator 1. Annex Table 4.A.1 lists established and emerging ocean-relevant sectors and industries.

Annex Figure 4.A.2. The ocean economy: A conceptual framework

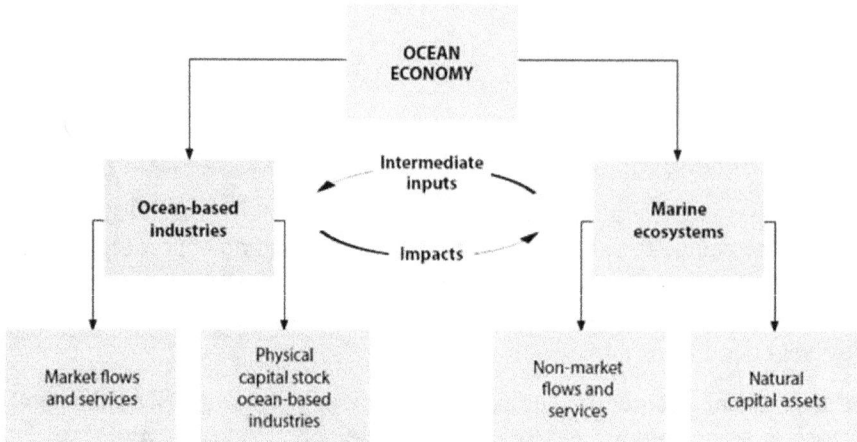

Source: OECD (2016[3]), *The Ocean Economy in 2030*, https://dx.doi.org/10.1787/9789264251724-en.

Annex Table 4.A.1. Emerging and established sectors and industries of the ocean economy

Established sectors	Emerging sectors
Capture fisheries and seafood processing	Marine aquaculture
Shipping and ports	Deep-water and ultra deep-water oil and gas
Shipbuilding and repair	Offshore wind energy
Offshore oil and gas (shallow water)	Ocean renewable energy
Marine manufacturing and construction	Marine and seabed mining
Maritime and coastal tourism	Maritime safety and surveillance
Marine business services	Marine biotechnology
Marine research and development and education	High-tech marine products and services
Dredging	

Source: OECD (2016[3]), The Ocean Economy in 2030, https://dx.doi.org/10.1787/9789264251724-en.

ODA to ocean-based industries and ecosystems provides a basis for understanding how much support development partners are channeling towards the ocean economy and which ocean-related sectors and industries are prioritised.

To calculate Indicator 1, the following steps are applied:

1. The OECD Creditor Reporting System (CRS) data fields of Project Title, Sector, Purpose Name, Short Project Description and Long Project Description are merged in an English-translated version of the database. In this way, the process can take advantage of the full informative content in the database; it also increases the likelihood that all ocean-relevant projects will be captured.

2. CRS projects in landlocked countries and specific sectors that were assessed as non-relevant to the ocean (for example, Basic Health, General Budget Support, Development Food Aid), are excluded. All projects associated with rivers, wetlands and lakes are also excluded to prevent the possible identification of projects that target freshwater bodies.

3. Using ocean-relevant keywords (Annex Table 4.A.2.), a keyword search is run to identify the subset of the CRS database containing only ocean-related activities. If a project contains at least one of the keywords (or partial words) shown in Annex Table 4.A.2. , it is classified as ocean-relevant and incorporated in the estimate.

Annex Table 4.A.2. Ocean and ocean economy keywords used for Indicator 1

Ocean economy keywords	ocean , oceans , oceans., ocean., \\(oceans\\), \\(ocean\\), seabird, grouper, red snapper, oceanograph, oceanic, gulf, oceanfront, oceangoing, ocean-going, algae, oceanarium, coral, mangrove, wave energy, fish, coastal, boat, marin, maritim, mpa, unclos , sea , seas , seawall, seabed, seagrass, seafood, sealife, seashore, seaport, seaboard, seawater, sea-, seashell, seaside, sargassum, submarin, sub-marin, seabeach, seaweed, seasalt, desalini, problue, wharf , seagull, offshore, vessel, problue, pro blue, pro-blue, deep water, international water, dredg, blue abadi, blue action fund, estuarine, estuary, saltwater, reef, blue economy, ship , ships , green shipping, shipping industry , shipping sector, shipbuilding, aquaculture, mariculture, cruise , naval , reef, coastal tourism, marine tourism, maritime tourism, tidal , beach , beaches , port , ports , harbour , harbor , tuna , shark, pelagic , reef , whale, bivalve, mussel, dolphin, salt marsh, oyster, shrimp , prawn , turtle, coral, crab, lionfish, blue action fund, problue, blue carbon, caribbean biodiversity fund, cockle, clam

Source: Authors.

ODA in support of the sustainable ocean economy

ODA figures on the sustainable ocean economy make it possible to quantify how much ODA is allocated towards ocean-related industries and activities that effectively integrate sustainability and how much towards specific actions to increase the resilience and the health of marine ecosystems. These figures also make it possible to determine the specific sub-sectors that are being prioritised within the sustainable ocean economy (e.g. sustainable fisheries, resilience, sustainable coastal tourism, etc.) and across providers and developing countries, and to see the range of instruments used.

Within the ocean economy ODA sample generated through Indicator 1, Indicator 2 captures only those projects that support specifically the conservation and/or sustainable use of marine and ocean resources. Indicator 2 identifies projects that are self-reportedly sustainable, in that are either reported as contributing to the Rio Conventions (i.e. marked through one of the Rio Markers for climate mitigation, climate adaptation and biodiversity); reported through the Environment Marker; or reported as contributing to environmental sustainability and resilience.

To calculate Indicator 2, the following steps are applied:

1. For bilateral donors and international organisations, Indicator 2 selects those projects among those in Indicator 1 that are marked through the Rio Markers for climate adaptation, climate mitigation and/or biodiversity and those that are marked through the Environment Marker.

2. While not all multilateral donors apply Rio Markers to their outflows, Multilateral Development Banks (MDBs) report on components of climate finance based on the MDB joint reporting approach. To make the estimate more homogeneous with the estimate for bilateral donors, projects identified as ocean-relevant and where banks report a Rio Marker component, the total amount is included in the estimate for Indicator 2.

3. In addition to screening them for markers, projects are screened through an additional keyword search aimed at identifying projects that address specific key areas for promoting sustainability and resilience. The keyword search was included for two reasons. First, the application of Rio Markers and the Environment Markers alone would lead to the exclusion of projects intended to build resilience through disaster risk preparedness, for example, or those supporting research to improve ocean health (SDG 14.A). Second, several projects in the CRS are not screened for Rio Markers and their sustainability can only be assessed by analysing the self-reported project description. The list of keywords used in this step is presented in Annex Table 4.A.3..

Annex Table 4.A.3. Keywords and filters used for Indicator 2

Markers	Environment, biodiversity, climate adaptation, climate mitigation
Sustainability keywords	sustainab , unclos , acidifica , mangrove , marine pollution , marine litter , marine debris , ocean debris , ocean pollut , ocean litter , ocean plastic , marine plastic , resilien , ecotouris , eco-touris , sargassum , conserv , marine protection , eutriphication , oxygen , ocean protection , mitigat , restoration , restore , restoring , sequestration , adapt , biodiversity , bio-diversity , biological diversity , bio-logical diversity , wildlife , waste , unclos , waste- , preserv , blue carbon , research , scientifi , science , mpa , protected area , renewable , green shipping , marine energy , wind energy , wave energy , tidal energy , biosphere , bio-sphere , bio sphere , offshore wind , off-shore wind , environment , disaster risk reduction , DRR , response preparedness , ecosystem , eco-system , nature-based solutions , blue carbon , carbon sink , SDG 4 , iuu , \\(iuu\\) , ocean health , habitat , over-fishing , overfishing , oceanograph , adaptation , sea level rise , sea-level rise , slr , \\(slr\\) , climate change , coral , turtle

Source: Authors.

ODA in support of land-based activities that positively affect the ocean

Indicator 3 makes it possible to quantify ODA in support of land-based activities that have a positive impact on the health of the ocean, such as actions to improve waste management and reduce agricultural runoff that pollutes the sea. This is a measure of indirect actions that are beneficial to the ocean and are thus considered as part of ocean-relevant ODA.

To calculate Indicator 3, the following steps are applied:

- The CRS data fields Project Title, Sector, Purpose Name, Short Project Description and Long Project Description are merged in an English-translated version of the database. In this way, the process can take advantage of the full informative content in the database; it also increases the likelihood that all ocean-relevant projects will be captured. CRS projects in landlocked countries are excluded, as are the sectors that were not considered relevant for the purpose of the indicator. All projects captured in Indicator 1 are also excluded to prevent overlap between ocean-based and land-based projects.
- Using relevant keywords (Annex Table 4.A.4), a keyword search is run to identify the sub-set of the CRS database containing only land-based, ocean-related activities
- If a project contains at least one of the keywords in Annex Table 4.A.4, it is classified as land-based, ocean-relevant activity and incorporated in the estimate.

Annex Table 4.A.4. Keywords used for Indicator 3

Land-based activities keywords	water treatment , agricultural runoff , waste , wastewater , sewage , marine pollution , marine litter , marine debris , ocean debris , ocean pollut , industrial runoff , ocean litter , ocean plastic , marine plastic , sewage , water resources conservation , river basins development , nutrient pollution

Notes

[1] The countries are Australia, Canada, Chile, Fiji, Ghana, Indonesia, Jamaica, Japan, Kenya, Mexico, Namibia, Norway, Palau and Portugal. The HLP is supported by the UN Secretary-General's Special Envoy for the Ocean.

[2] The OECD Creditor Reporting System contains no marker or immediate way to retrieve data on ODA for the ocean. Our Shared Seas has made an attempt to provide estimates on ocean-related ODA, but its estimates do not aim to assess the share of ODA that goes towards sustainable activities. The limitations of this methodology and how it differs from the OECD ocean-relevant methodology are discussed in Annex 4. A.

[3] In 2013-2018 LMICs received 42%, LDCs 41%, and UMICs 16% of global ODA.

[4] This figure refers to 17 of the top 20 recipients of ocean economy ODA. Integration of sustainability in this ODA is much higher than the average for three of the top 20 recipients, namely: Bangladesh, India and Madagascar.

[5] This figure refers to the average ratio of EEZ to land mass in the 34 ODA-eligible SIDS. This ratio is highest for Tuvalu (EEZ exceeds its land mass by 28 838 times), followed by Nauru (EEZ exceeds its land mass by 14 689 times).

[6] This estimate considers the Cook Islands and the Seychelles as SIDS, although they both recently graduated from being ODA-eligible.

5 How development co-operation can help re-orient private finance and investments towards a more sustainable ocean economy

This chapter focuses on how development co-operation can support scaling up private finance for sustainable ocean economies and scale down and re-orient finance that now flows to ocean practices and activities that are harmful and unsustainable. It starts by quantifying the private finance mobilised for the sustainable ocean economy by official development assistance through leveraging instruments such as guarantees, co-financing schemes, etc. It then identifies a range of new financial instruments – including blue bonds, debt-for-nature swaps, new blue carbon schemes and ocean risks management tools – developed with the support of the development co-operation community. It discusses the current market size of these new instruments, as well as main challenges and opportunities for replicating them and scaling them up. The chapter closes with a set of suggestions for re-focusing financial flows away from unsustainable ocean activities and streamline investment sustainability.

The need to increase resources goes hand-in-hand with the need to re-orient private finance away from harmful activities

The systemic, transformational and urgent changes needed to achieve sustainable ocean economies require action by all actors, public and private. The private sector is essential to this transition – both for increasing the scale of sustainable investments for healthy and productive oceans and for crafting new, more environmentally and socially sustainable business models and products in ocean-based sectors. But this change will not happen on its own. Public policies and financing need to be proactive (Kattel et al., 2018[1]), consistent and commensurate with the need to align private finance to the sustainability imperative of the global ocean economy, and actively co-creating markets and tilting the playing field in the direction of sustainability for shared prosperity.

In the same vein, development co-operation can be used to directly fund sustainable investments (as explored in Chapter 4) and to help re-direct private finance towards sustainable businesses and activities. Development co-operation can help increase the availability of finance for sustainable investments through support towards innovative financial instruments and other leveraging instruments that can mobilise a broader array of public and private resources for the sustainable ocean economy. However, these flows may be a drop in the ocean without greater efforts to curb and re-orient the financial flows currently fuelling destructive practices that often have the largest impacts on developing countries' fish populations, coasts and tourism, food security, and livelihoods. Therefore, development partners have a critical role in supporting policies, regulations and financial levers to divert finance from harmful and unsustainable practices and to ensure that sustainability is integrated in traditional financial services and investments, in financial markets (e.g. stocks and bonds), and in credit markets (e.g. loans or bonds).

Blended finance can help mobilise additional finance towards sustainable ocean economies in developing countries. But not all blended finance is quality blended finance. The OECD posits that the "quality" of blended finance means achieving the core mission of the Sustainable Development Goals to leave no one behind (OECD, 2018[2]). Therefore, a proven development contribution is essential. Also critical is the temporal dimension. Blended finance should aim to change the market and achieve scale and so, rather than being a permanent feature in private investments, it should be a time-bound intervention that is part of a broad, ambitious and strategic approach for mobilising additional resources through ODA. Beyond attracting commercial capital in a transaction, the ambition of blended finance is to be catalytic, i.e. to spur the replication of similar projects via demonstration and build sustainable markets and products.

This chapter identifies and discusses the range of ODA leveraging financial instruments – grants, standard loans and guarantees, and syndicated loans – used to mobilise private finance for sustainable ocean economies and quantifies how much private finance has been mobilised through these instruments. It reviews the range of new financial instruments being developed with support from development co-operation for mobilising private capital for sustainable ocean economies, including blue bonds, debt-for-ocean swaps, new blue carbon schemes, and ocean risks management tools such as marine and coral reef insurance. The chapter analyses the current market size of these instruments and discusses challenges and opportunities for replicating them and scaling them up. Finally, the chapter explores how development partners can support the regulations and policy levers that are needed to phase out financing for harmful practices and redirect financing towards sustainable uses while contributing to international policy coherence for sustainable development.

Quantifying private finance mobilised for the sustainable ocean economy through development finance instruments

In the 2013-17 period[1] development finance mobilised a total of USD 2.92 billion of private finance in support of ocean-related projects, equivalent to USD 585 million on average a year. This includes private finance mobilised for ocean-based industries and ecosystems (USD 1.3 billion, or 43%) as well as private finance mobilised for land-based activities that reduce negative impacts on ocean, such as waste management, sanitation and water treatment (USD 1.7 billion, or 57%) (Figure 5.2). These flows fluctuated significantly through the period and overall, point to a slight downward trend (-11% over the period).

To mobilise private finance for the ocean, development partners employed a number of leveraging official development assistance (ODA) instruments, including standard grants and loans, guarantees, direct investments in companies (i.e. equity), collective investment vehicles, credit lines, syndicated loans, and simple co-financing schemes. Some of these instruments were used to improve the viability of commercial investments and to make projects more attractive by de-risking investments or helping to structure returns through new and emerging blended finance arrangements. Development partners also often provide technical assistance during the project preparation phase, crucial for the overall success of the project.

Providers of development co-operation used these instruments in different contexts and for different ends, showing that these lend themselves to different purposes and that they are not interchangeable. For land-based activities, development finance was able to mobilise the greatest volumes of private finance using guarantees, chiefly employed in support of water and sanitation interventions. In the case of ocean-based industries, the largest amounts of private finance were mobilised through direct investments in companies and special purpose vehicles (SPVs) as well as through syndicated loans. These instruments were used primarily for water transport projects, such as ports. In the fisheries sector, 70% of private finance was mobilised through guarantees, while direct investments and co-financing schemes mobilised smaller amounts (26% and 4%, respectively). For coastal and marine protection, private finance was exclusively mobilised through co-financing schemes.

Globally, upper middle-income countries (UMICs) benefited the most in 2013-17 from the leveraging effect of development finance instruments, receiving 39% of the overall amounts mobilised, mainly in South America (Figure 5.2). However, the picture is different when land-based activities and projects in ocean-based industries and conservation are separated out. For land-based activities alone, the largest volumes of private finance were mobilised in least developed countries (LDCs), primarily through guarantees facilitating water and sanitation projects. Private finance for ocean-based industries and ocean and coastal conservation was mobilised primarily in UMICs, where these investments mainly targeted ocean-based industries (68%), largely for water transport, with the remaining 32% directed to curbing the negative impacts from land-based activities, mainly through investments in sewage and disposal systems.

For LMICs and LDCs, investments in land-based, ocean-related sectors, and particularly for projects on waste management, water supply and sanitation, attracted the vast majority of the mobilised private capital, accounting for 89.8% in LMICs and 99.9% in LDCs. For the two country groups, guarantees were the largest mobilisation tool, 98.8% across LDC countries and 50.5% in LMICs. For LMICs, however, private finance was mobilised through a wider array of financial instruments including direct investments in companies and SPVs, credit lines, resource pooling mechanisms (collective investment vehicles), and simple co-financing arrangements summarises how different leveraging development finance instruments were used to promote the conservation and sustainable use of the ocean across countries. Figure 5.3 provides a summary by leveraging instrument to show how much finance each instrument mobilised, in what countries and for what purposes.

Figure 5.1. ODA mobilised private finance for the sustainable ocean economy largely in upper-middle income countries

Source: Authors' calculations based on OECD DAC (2020[3]), Amounts mobilised from the private sector for development, http://www.oecd.org/development/stats/mobilisation.htm

StatLink https://doi.org/10.1787/888934159658

Figure 5.2. Amounts of private finance mobilised for the ocean through official development assistance

Source: Authors' calculations based on OECD DAC (2020[3]), Amounts mobilised from the private sector for development, http://www.oecd.org/development/stats/mobilisation.htm

StatLink https://doi.org/10.1787/888934159677

Figure 5.3. Key facts about private finance mobilised for the ocean, by ODA instrument

Cumulative figures for 2013-17

	Total volume of private finance mobilised (million)	Income groups	Number of transactions	Average size of transaction	Share of private finance mobilised for the ocean economy	Share of private finance mobilised for ocean relevant land based activities	Top sector
Guarantees	USD 1,258.1	LMICs (40%), LDCs (57%), UMICs (3%)	25	USD 50.3 million	7%	93%	 Water and sanitation
Syndicated Loans	USD 905.3	UMICs (88%), LMICs (11%), LDCs (1%)	12	USD 75.4 million	64%	36%	 Water transport
Direct Investment	USD 638.2	UMICs (51%), LMICs (48%), LDCs (1%)	31	USD 20 million	96%	4%	 Water transport
Credit lines	USD 82.6	LMICs (100%)	3	USD 27.5 million	16%	84%	 Ocean related environmental policy
Cofinancing	USD 42.3	LMICs (47%), LDCs (45%), UMICs (8%)	102	USD 0.1 million	6%	94%	 Water and sanitation
Share in collective investment vehicles	USD 30.3	LMICs (100%)	3	USD 10 million		100%	 Water and sanitation

Water and sanitation  Water transport  Ocean related environmental policy

Source: Authors' calculations based on OECD DAC (2020[3]), Amounts mobilised from the private sector for development, http://www.oecd.org/development/stats/mobilisation.htm

StatLink https://doi.org/10.1787/888934159696

Innovative financial instruments for the sustainable ocean economy

In addition to using grants and other established financial instruments to crowd in private finance, development partners are supporting the development of new and innovative financial instruments to attract a broader set of resources for the conservation and sustainable use of the ocean in developing countries. Development partners have supported the creation and implementation of new financial instruments and products in a variety of ways: through technical assistance, by absorbing the costs of the development phase of these new instruments, by supporting the identification of a pipeline of bankable projects, and providing concessional finance to improve the viability and attractiveness of commercial investments by bringing down the risk-adjusted financing rate. The strong leadership and commitment of developing country governments was critical for testing and developing several of these new financial products and approaches, which they saw as opportunities to enhance fiscal space and invest in their ocean resources.

The following sections provide a brief overview of innovative financial instruments and approaches supported by development finance and implemented for financing the conservation and sustainable use of the ocean. They explore the current market size of these new financial instruments, discuss the opportunities and challenges for scaling them up, and highlight how development co-operation providers could support their replication. Innovative financial instruments are clustered in four categories:

- new debt instruments, such as the first sovereign blue bond and the debt-for-ocean swap implemented by the Seychelles
- blue carbon schemes
- risk management tools, such as insurance schemes for marine and coastal resources where donor funding is used to lower the risk premium
- impact investing for the ocean, including through new products and funds such as the Althelia Sustainable Ocean Fund, Encourage Capital and the Meloy Fund.

Blue bonds

Current blue bond market

Blue bonds are a relatively new type of sustainability bonds issued to finance projects relating to the conservation and sustainable use of the ocean and the transition towards a sustainable ocean economy (World Bank, 2018[4]). Operating similarly to other debt instruments, blue bonds provide capital to issuers who repay the debt with interest over time. Bond investors are paid a fixed interest rate (coupon) on a fixed schedule and will be returned their initial investment (principal) upon maturity of the bond.

The first blue bond was launched by the government of the Seychelles in October 2018 for USD 15 million with a maturity of 10 years and a coupon (annual interest payment) of 6.5%. In January 2019, the Nordic Investment Bank issued a Swedish kroner (SEK) 2-billion (USD 200 million) blue bond to protect and rehabilitate the Baltic Sea. The proceeds from the Seychelles blue bond are used to support the expansion of marine protected areas, to improve governance of priority fisheries and for the development of the Seychelles' blue economy. Grants and loans for sustainable projects are provided through the Blue Grants Fund and Blue Investment Fund, managed respectively by the Seychelles Conservation and Climate Adaptation Trust (SeyCCAT) and the Development Bank of Seychelles. In its design, the Seychelles blue bond resembles many features of a green bond. However, it was not issued as a green bond, as experts involved in the transaction considered that existing taxonomies of "green projects" do not capture all facets of preserving and protecting marine ecosystems targeted by the blue bond (Roth, Thiele and von Unger, 2019[5]). It is true that criteria or definitions of eligible green projects for green bonds mainly revolve around emissions reduction and energy efficiency, while the scope of sustainable ocean-based activities is substantially larger (United Nations Global Compact, 2020[6]).

Several green or sustainable bonds also were issued recently that incorporate blue elements. In 2019, the World Bank issued a USD 10-million sustainable development bond, with proceeds focusing on plastic waste reduction in the ocean and the sustainable use of marine resources in developing countries including via scientific research and regulatory reform (Morgan Stanley Institute for Sustainable Investing, 2019[7]). In 2017, Fiji was the first developing country to issue a green bond, the Fiji International Finance Corporation Green Bond, comprising elements relating to blue natural capital. The size of this sovereign bond was 100 million Fijian dollars, equivalent to USD 50 million. In March 2018, the government of Indonesia issued the first sovereign Green Sukuk, a green Islamic bond. The issuance attracted conventional, Islamic and green investors and was oversubscribed, signalling the growing demand for sustainable and responsible investment. Proceedings financed a range of projects including the replacement of fossil fuel-derived electricity with solar PV-based batteries for sea navigation facilities such as lighthouses (Indonesia Ministry of Finance, 2019[8]). The Climate Bond Initiative also has developed criteria for marine renewable energy to be incorporated in the Climate Bonds Standard and Certification

Scheme, which is a labelling scheme for bonds and loans used globally by bond issuers, governments, investors and the financial markets to prioritise investments that contribute to addressing climate change. The marine renewable energy criteria have been designed to encompass both established and emerging marine renewable energy technologies and can be applied to offshore wind, offshore solar, wave power and tidal power.

Blue bonds are a financing opportunity not only for governments but also for corporations, which could use the blue bond market to leverage resources for investments aligned to the sustainable use of the ocean. While corporate blue bonds are still at the development stage, the United Nations (UN) Global Compact has developed a Blue Note to assess opportunities for using the environmental, social, governance bond market for corporations in search of commercial financing for ocean-related investments. The nine principles defined in the Blue Note identify criteria to ensure ocean health and productivity, ocean governance and engagement, and data and transparency and are a first step towards a taxonomy in this area. Signatory companies commit to being aware of their impact on the ocean, to revise their company strategy accordingly and to disclose their actions (United Nations Global Compact, 2020[6]).

Opportunities and challenges to scaling up blue bonds

To investors, blue bonds offer an opportunity to diversify their investment portfolios into products that generate a financial return and deliver environmental and socio-economic benefits linked to the ocean. A bond is a well-known financial instrument to the traditional investor who has not necessarily been active in the conservation or development finance areas and, therefore, it eliminates some of the fears and concerns around trying new or innovative and unproven investments. A bond furthermore comes with the security for investors of an independent credit rating, in a language that investors understand (Fritsch, 2020[9]). For issuers, the blue bond helps raise resources with a small interest discount obtained against the commitment to use funds for sustainable ocean-related projects. Blue bonds are therefore attractive financial instruments to both investors and issuers and have the potential to harness capital market resources for marine conservation and sustainable ocean-based activities.

More blue bonds are expected in the coming years as the opportunities from expanding ocean-based industries and the need for building resilience become more apparent (Roth, Thiele and von Unger, 2019[5]). Several countries are actively exploring the feasibility of blue bonds, especially small island developing states (SIDS) that have a big stake in a more sustainable ocean economy but where perceived high investment risks and small size of operations significantly constrain investors' appetite. Blue bonds are under consideration by Cabo Verde, Caribbean countries (Caribbean Development Bank, 2018[10]) and the Pacific islands through a 'Pacific Ocean Bond (Walsh, 2018[11]). Table 5.1 presents an overview of existing and prospective blue bonds as well as bonds with 'blue element'.

The World Bank feasibly study for the issuance of a blue bond in Cabo Verde offers insights on opportunities and challenges that may apply to other countries. In terms of benefits, the study noted that the blue bond would raise the profile of Cabo Verde on the blue economy globally, potentially attracting other sources of financing and partnerships. According to the World Bank, the blue bond also would contribute to building domestic capacities to diversify borrowing sources, a crucial step after Cabo Verde graduated from LDC status and the consequent reduction in its access to concessional lending. While the first sovereign blue bond issued by the Seychelles was strongly focused on sustainable fisheries, the potential exists for investments in a broader range of ocean-based sectors, and the World Bank found that the blue bond of Cabo Verde would contribute to further refining blue bond criteria beyond fisheries.

The feasibility study, however, pointed also to a number of challenges and constraints. For instance, although the blue bond would present credit enhancement features that would lower the cost of a blue bond in comparison to a standard bond through the use of guarantees and other development finance instruments, the bond would still represent a source of financing with a higher interest rate and a shorter maturity than concessional finance. More importantly, the blue bond would further increase the indebtedness of Cabo Verde, one of the most heavily indebted countries of sub-Saharan Africa. Finally, the World Bank currently has no country programme in Cabo Verde in which to anchor this initiative and, as is the case with other SIDS, Cabo Verde is not a priority country for the development co-operation of several providers. Therefore, the country effectively faces limited access to concessional finance, due to its middle-income status and the high transaction costs associated with providing support to a small country.

Overall, the current interest in issuing blue bonds in various countries suggests they could be a significant source of financing for sustainable ocean economies, but some prerequisites need to be met. These include:

- Countries need a plan. This aspect was crucial to the issuance of the Seychelles blue bond. A plan could be a clear mission-oriented strategy for the sustainable ocean economy based on evidence and adopting a coherent cross-sectoral approach.
- Blue bonds need to be one in an array of financing instruments. Given the current size of blue bonds, they are unlikely to be the main or sole financing instrument for promoting a sustainable ocean economy. Ideally countries would develop a financing strategy aiming to achieve a balanced mix of instruments that takes into consideration the pros and cons of various sources and instruments and maintain the country on a path of debt sustainability.
- A pipeline of blue bond projects is needed. This pipeline of projects needs to be aligned with the national, mission-oriented strategy for the sustainable ocean economy and effectively promote the conservation and sustainable use of the ocean. The development of appropriate projects with identified returns and robust assessments of their positive impacts on marine and coastal ecosystems is identified as a critical gap to date.

The role of development partners in supporting the scaling up of blue bonds and ensuring they are sustainable

Support from development co-operation has been key to the emergence of blue bonds, taking several forms. Development co-operation providers extended technical assistance and provided expertise for structuring the financial product. This was critical for the Seychelles Blue Bond, which required sophisticated financial engineering and about 18 months of preparatory work. Support from development co-operation providers also included the provision of concessional finance for credit enhancement. For the Seychelles Blue Bond this consisted of a USD 10-million finance package from the World Bank and the Global Environment Facility (GEF), consisting of a USD 5-million guarantee from the International Bank for Reconstruction and Development (IBRD) and a USD 5-million grant from the GEF. The credit enhancement allowed for a reduction of the price of the bond by partially de-risking the investment for the investors and by reducing the effective interest rate for the Seychelles from 6.5% to 2.8% (World Bank, 2018[4]). Too small to be traded on an exchange, the Seychelles Blue Bond was sold in a private placement to three impact investors based in the United States.[2]

Development partners have an important role in further supporting countries to develop and use blue bonds effectively as an innovative financial instrument for the conservation and sustainable use of the ocean. Their contributions can include:

- supporting a clear and coherent strategy for the sustainable ocean economy and a financing plan that blue bonds will be part of
- supporting the development of a pipeline of projects that will help translate the strategy into investments that effectively promote the conservation and sustainable use of the ocean and guard against so-called blue-washing
- contributing to meeting the costs of bonds design and providing technical assistance to enhance domestic capacities to manage complex financial structuring and potential implications, especially

for countries, such as Cabo Verde, that have no or limited experience borrowing in the global capital markets.

Table 5.1. Overview of blue bonds and other bonds with blue elements

	Bond	Issuing Institution	Transaction size	Use of proceeds
Blue bonds	Seychelles' Sovereign Blue Bond	Government of Seychelles	USD 15 million	Management of marine protected areas and sustainable fisheries, including the creation of more value for seafood products
	Pacific Ocean Bond	Pacific islands	To be determined	To be determined
Green and sustainable bonds with blue elements	World Bank's Sustainable Development Bond	World Bank	USD 10 million	Plastic waste reduction in oceans sustainable use of marine resources; scientific research; policy and regulatory reform in developing countries
	Fiji Green Bond	Government of Fiji	USD 50 million	Renewable energy; water and energy efficiency; clean transport; wastewater management and sustainable agriculture to reduce fertiliser runoff into the ocean.
	Indonesia Green Sukuk	Government of Indonesia	USD 1.25 billion	Improvement of solid waste management system at city and regional scale; Replacement of aid for sea navigation facilities (e.g. use of renewable energy for lighthouses).

Source: Authors

Debt-for-ocean swap

Current debt-for-ocean transactions

In 2015, the government of Seychelles and its Paris Club[3] creditors implemented a debt-for-nature swap that allowed the country to reduce immediate debt burdens while also increasing resources targeted towards ocean and climate action. Although debt conversion has been used in the past for conservation purposes (Asiedu-Akrofi, 1991[12]), the Seychelles debt-for-nature swap was the first to use proceeds for marine conservation and adaptation and the first to include a private philanthropic institution for raising the funds to purchase the sovereign debt (complementing The Nature Conservancy's capital) (Silver and Campbell, 2018[13]). The transaction also marks the first time Paris Club creditors have supported a debt buyback designed to benefit the environment, and creditor participation in the agreement is the highest ever achieved in a buyback reached through the Paris Club's market-based window (Convergence, 2017[14]).

The debt-for-ocean swap covered a portion of the Seychelles' total debt amounting to USD 20.2 million. The creation of SeyCCAT to be in charge of managing the loan transactions and providing funding for the conservation and adaptation initiatives was a crucial part of the structuring process (Silver and Campbell, 2018[13]). SeyCCAT was capitalised by a debt conversion structure put together by The Nature Conservancy comprising USD 5 million in grants and a USD 15.2-million loan from NatureVest. With these resources, SeyCCAT lends the government of Seychelles USD 20.2 million to repay its debt from official creditors at more favourable terms, particularly longer tenors and payments in local currency. In turn, the government issued notes to SeyCCAT committing to pay back USD 15.2 million at a 3% rate over 10 years and USD 6.4 million at a 3% rate over 20 years. SeyCCAT will use the proceeds of these notes to repay the USD 15.2-loan with The Nature Conservancy; fund activities related to marine and coastal conservation including strategies for ecosystem-based climate adaptation and disaster risk reduction, for a total of USD 5.6 million; and conduct USD 6.6-million capitalisation of the Fund for future programming.

As part of the debt-for-ocean swap, the government of Seychelles has increased its marine protected area to 30% of its territorial waters, and half of this area, equivalent to approximately 2 000 km² was defined as a no-take fishing zone. The country will also conduct a marine spatial plan for its entire exclusive economic zone and establish a permanent trust fund for marine conservation and climate adaptation activities (Convergence, 2017[14]; The Nature Conservancy, 2020[15]).

Opportunities and challenges for scaling up debt-for-ocean swaps

Debt-for-ocean swaps could be a useful tool for highly-indebted countries, especially SIDS with high debt burdens and vast ocean resources to protect and sustainably use. It is no coincidence that Antigua and Barbuda has explored a country-specific debt-for-ocean swap (Fuller et al., 2018[16]), as has Grenada, and that proposals for debt-for-ocean swaps of various kinds have been developed for small states and Caribbean SIDS. The World Bank proposed a debt-for-nature and resilience facility for small states; the Commonwealth Secretariat developed a similar proposal for a debt-for climate swap facility for small vulnerable economies (Caribbean Development Bank, 2018[10]); and the UN Economic Commission for Latin America and the Caribbean developed a proposal for a mechanism to use climate finance pledges to write down the high debt of Caribbean countries in exchange for investments in climate adaptation and mitigation (UN, 2016[17]). Building on the successful experience of the Seychelles, The Nature Conservancy is seeking to finance USD 1-billion of new debt conversions, or blue bonds, around the world.

To date, however, proposals have not translated into new ocean-for-debt swaps or facilities. Overall, debt-for-ocean and debt-for-nature swaps can be fairly complex and lengthy works of financial engineering. They are also often not positively welcomed by development partners on the grounds that they could create perverse incentives and induce moral hazard, favouring the accumulation of debt in view of a later cancellation (IMF, 2016[18]). Grenada initially explored a debt-for-ocean swap, but this become a less attractive option due to the improvement in its debt situation following successful debt restructuring efforts.

Convergence (Convergence, 2017[14]) points to a number of considerations to take into account for developing debt-for-ocean swaps similar to that of the Seychelles. The study highlights that the strong ownership and leadership of the Government of Seychelles was a critical element for the successful implementation of its debt-for-ocean swap, together with the interest from official creditors willing to sell debt owed by the Seychelles (Convergence, 2017[14]). The study points to additional factors that are turned into opportunities for development partner support in the next section.

The role of development co-operation in scaling up debt-for-ocean swaps

Under the Seychelles debt swap mechanism, providers used concessional finance to gradually write down the country's debt stock on the condition that funds otherwise used for debt service payments would be used for climate and ocean investments. To support other countries in structuring debt conversions to enhance financing of their sustainable ocean economies, development co-operation partners can provide the following:

- Technical assistance: Debt conversions are complex and require extensive analysis and negotiations. The debt structuring of the Seychelles debt conversion, for instance, took approximately four years.
- Early funding commitments: These can have a significant impact on pushing a debt conversion forward. In the case of the Seychelles, an early commitment of USD 1 million by one foundation gave the Seychelles government confidence that there was real funder interest for implementing the debt conversion.

- Creditor consent: Several bilateral development partners are Paris Club creditors, and in the case of the Seychelles, Belgium, France, Italy, Belgium, South Africa and the United Kingdom were among the Paris Club creditors agreeing to the debt conversion.

Blue carbon schemes

New carbon schemes approaches: blue carbon schemes

Because many natural ecosystems have the capacity to act as net carbon sinks, conserving and restoring them has been recognised as an important part of climate change mitigation. The UN Framework Convention on Climate Change (UNFCCC) has adopted policies to allow countries to account for gained and lost carbon emissions through land use change, both by including these emissions in national assessments and by providing mechanisms to fund and incentivise conservation projects to respond to this significant contributing factor of anthropogenic climate change.

More recently, studies have highlighted coastal ecosystems such as mangroves, salt marshes and sea grasses as significant carbon sinks that are able to contribute to climate mitigation. Although these ecosystems make up only 2% of global area, their capacity to sequester carbon dioxide is ten times more effective on a per area basis per year than the capacity of forests (Fourqurean et al., 2012[19]) and about twice as effective at storing carbon in their soil and biomass (Murray et al., 2011[20]). Restoration of vegetated coastal areas, or coastal blue carbon, contribute to climate change mitigation through increased carbon sequestration and storage of around 0.5% of current global emissions annually (IPCC, 2019[21]). Besides carbon sequestration, ocean and coastal ecosystems provide a multitude of other benefits, including storm protection and resilience, water quality, and biodiversity.

Coastal blue carbon ecosystems, however, have degraded at an alarming rate (Pendleton et al., 2012[22]): it has been estimated that one third of the totality has been lost over the past decades. Their degradation not only reduces their capacity to store carbon. It actually releases sequestered carbon into the atmosphere. Degradation of blue carbon systems is estimated to cause emissions of between 0.15 and 1.02 billion tonnes of carbon in the atmosphere each year; that is the equivalent of an amount ranging from the annual release from deforestation in the Amazon up to 6 times this annual release (Vanderklift et al., 2019[23]). Further, since the 1970s, the ocean has absorbed more than 90% of the excess heat in the climate system as well as excessive CO_2, resulting in increasing ocean acidification and loss of oxygen, with large environmental and socio-economic consequences (IPCC, 2019[21]).

To tackle the negative impacts of blue carbon ecosystems degradation, conservation and restoration projects can be developed to effectively contribute to climate change mitigation in ways that generate new and additional revenue streams for marine and coastal ecosystems. The first coastal blue carbon projects are currently being developed and are testing new approaches, methodologies and financing solutions (Wylie, Sutton-Grier and Moore, 2016[24]). Under the policy frameworks and trading allowances schemes between countries created by the UNFCCC, however, credits from coastal blue carbon such as mangroves restoration have hardly been traded (Herr et al., 2018[25]). Increasingly, however, stakeholders are turning blue carbon into tradeable assets on voluntary markets, and the international community is evaluating how these systems can be more effectively included in existing policy frameworks, including carbon financing mechanisms such as Reducing Emissions from Deforestation and Forest Degradation (REDD+) and other UNFCCC mechanisms (Herr, Pidgeon and Laffoley, 2012[26]).

Opportunities and challenges for scaling up blue carbon schemes

While climate mitigation and ocean-based industries and livelihoods stand to benefit greatly from the conservation and restoration of blue carbon ecosystems, scaling up blue carbon schemes will hinge on addressing some critical challenges and opportunities linked to these schemes. Key opportunities for successful and replicable blue carbon scheme include:

- Incorporating livelihoods aspects: Successful blue carbon projects are those that involve local communities from the very start of the project design and throughout all stages of planning and implementation (Wylie, Sutton-Grier and Moore, 2016[24]). These projects incorporate the needs of local communities, identify direct benefits for the community, manage trade-offs and ensure that increased protection in one area does not translate into greater exploitation in another.
- Exploring the potential of blue carbon ecosystems beyond mangroves: Blue carbon projects have so far focused primarily on mangroves, likely because of their abundance in tropical areas where projects have been developed, and because UNFCCC mechanisms, such as REDD+, only include mangrove ecosystems. However, other blue carbon ecosystems, such as marshes and sea grass, present a high rate of carbon sequestration that exceeds that of terrestrial forests. These blue carbon ecosystems could effectively be included in future projects.
- Including carbon stored within the soil: Only few projects account for carbon stored within the soil of blue carbon ecosystems, although this stores 95-99% of total carbon stocks for tidal marshes and sea grasses and 60-80% for mangroves (Murray et al., 2010[27]). This suggests that current projects are below their financial potential.
- Incorporating the resilience value of marine and coastal ecosystems besides mitigation value: Coastal wetlands not only contribute to the carbon cycle but also protect coastline by attenuating wave energy and providing protection against storms damage (Spalding et al., 2014[28]). Incorporating this value in the design and implementation of the blue carbon would increase the total value of the blue carbon project.
- Adopting the most suitable financing mechanism: So far, voluntary markets and alternative financial mechanisms have proved to be best suited for small, community-based projects, mainly because their costs are lower than those of UNFCCC mechanisms. This may change as the international community gains experience in effectively integrating blue carbon projects into various UNFCCC mechanisms (Wylie, Sutton-Grier and Moore, 2016[24]). Further, the new co-operative approach envisioned in the Paris Agreement to achieve climate change pledges allows countries to identify and implement interventions bilaterally, potentially including coastal blue carbon, and this could make blue carbon projects more replicable and able to attract new and larger finance flows (Herr et al., 2018[25]).

The role of development co-operation in supporting the scaling up of blue carbon schemes

Most blue carbon projects in developing countries are supported in one way or another by development co-operation providers. They can thus an important role to ensure that the value of these projects for local communities and the environment, as well as their financial value, are maximised. This will mean supporting countries to explore some of the opportunities discussed in this chapter.

Recognising the social, economic and environmental benefits of blue carbon initiatives (also referred to as co-benefits), in 2017 the Australian government launched a USD 6-million allocation to support efforts to protect and manage coastal blue carbon ecosystems in the Pacific, in partnership with Fiji, Papua New Guinea and other Pacific countries, as well as regional and private stakeholders. This funding will strengthen expertise and enhance data availability for conducting blue carbon projects in the Pacific and support their integration into national greenhouse gas accounting and climate policies. It will also encourage public and private sector investment through innovative financing approaches to blue carbon projects.

To explore how to structure a system of credits incorporating a market value for the resilience services provided by marine and coastal ecosystems, in 2019 the Nature Conservancy, with support of the international insurance company AXA XL, launched the "Blue Carbon Resilience Credit" initiative. Considering the resilience value of marine and coastal ecosystems would allow companies to reduce their carbon footprint and to contribute to maintaining the services these ecosystems provide, including protecting coastal populations from natural hazards, and in this way avoiding the costs of reconstruction. The methodology for determining the value of the resilience services provided by each coastal ecosystem and the structure of the credit system is currently under development.

New ocean-relevant risk management tools

Current market

Extreme weather events and other natural hazards are expected to increase in the near future (IPCC, 2019[21]). These events can wipe out decades of development gains overnight and have devastating and life-changing consequences especially for already vulnerable communities. According to the UN Office for Disaster Risk Reduction, direct economic losses from climate-related disasters rose 151% over the last 20 years. In the last 10 years alone, insurers have paid out around USD 300 billion following storm damage to coastal regions, and the costs to governments and taxpayers have been far higher (UN, 2015[29]). By 2050, estimates suggest that the global community will face annual costs of USD 1 trillion from the combined effects of rising sea levels and extreme weather events on coastlines.

It is therefore critical to address ocean risks and close the gap between the level of insurance in place to cover ocean risks and the actual cost to businesses, governments and people of rebuilding and recovering from disasters. To date, two innovative initiatives have emerged: (i) Mexico's parametric coral reef insurance, a country-specific scheme developed with support from The Nature Conservancy, the United Nations Development Programme and other development partners, and (ii) the Ocean Risk and Resilience Action Alliance (ORRAA), established under the auspices of the Canadian Presidency of the G7.

In 2018, the government of the state of Quintana Roo in Mexico purchased a parametric insurance product that would offer up to USD 3.8 million to cover hurricane-related damage to coral reefs. The area is in fact highly vulnerable, as shown by 2005 Hurricane Emily, which generated damages for USD 8 billion in the Cancun and Puerto Morelos areas. The parametric insurance is provided by Mexico-based insurer Afirme Seguros Grupo Financiero SA de CV, and will be triggered if wind speeds above 100 knots are registered within the covered area, with a payout split of 50% for reefs and 50% for beaches (Secretaria de Ecologia y Medio Ambiente Quintana Roo, 2019[30]; The Nature Conservancy, 2020[31]).

To finance the insurance policy and promote the conservation of coastal areas, the Quintana Roo State government created the Coastal Zone Management Trust, financed by taxes collected from the tourism industry. This shows the close connection between the tourism sector and marine and coastal ecosystems and how tourism sector revenues can be harnessed for the conservation of marine and coastal ecosystems. Reefs sustain the tourism industry of Quintana Roo by providing coastal protection against storms, including by reducing beach erosion, but extreme storms put the reefs at risk. The Trust will direct conservation investments for the maintenance and repair of the reef and beaches while managing the insurance payout and ensuring conservation goals are achieved.

The ORRAA is another example of a new initiative in support of ocean-relevant risk management tools. It was launched in 2019 the with support from the government of Canada to foster collaboration among governments, financial institutions, the insurance industry, environmental organisations and other stakeholders with the aim of creating innovative finance solutions that build resilience to ocean risk in the regions that need it the most, including small island states. It is designed to drive USD 500 million of investment into coastal natural capital by 2030.

The Alliance focuses on three main priorities. The first is to develop innovative, risk-adjusted and scalable products that help drive investment into coastal natural capital and increase resilience while delivering a return on investment. The products include nature-based insurance and micro-insurance, carbon credit initiatives, micro-finance, green and/ or blue bonds, and resilience bonds. The second priority is to accelerate the research and data collection needed to better analyse, model and manage ocean risk. In collaboration with several partners, AXA XL is leading the development of an Ocean Risk Index to develop potential scenario analyses of the implications of sea-level rise and habitat degradation, with the benefit of helping motivate the protection of ecosystems. The third ORRAA priority is to raise awareness and public understanding of ocean risks and inform and advance ocean risk policy among governments and the private sector.

Opportunities and challenges for scaling up ocean-relevant risk management tools

Greater expected impacts from extreme weather events call for innovative products that incorporate and manage ocean risks and that can effectively generate adequate resources to preserve and restore natural capital essential to resilience. The two examples are promising, but they remain isolated initiatives, suggesting there is scope for replication and scaling-up.

The Mexican insurance scheme for coral reefs, in particular, has high potential for replicability in coral regions and other regions where ecosystems are threatened by extreme natural hazards. Because of how it is structured, this parametric insurance effectively provides a new source of conservation finance. It demonstrates that natural capital can be effectively integrated in emergency response planning and implementation. Due to its characteristics, it gives countries the incentive to undertake continuous preventive work and enhance capacities to assess climate risks to coral reefs and different ecosystems. Challenges to replicating and scaling up these innovative insurance schemes include the availability of adequate data and modelling to evaluate the benefits from marine and coastal ecosystems (Reguero et al., 2019[32]). Further, instruments such as the Mexican insurance scheme would only work in highly touristic areas. It proves more difficult to apply such instruments in areas where resilience building is needed but that are not as attractive for tourists. Therefore, the design of coastal risk management policies and tools should take into account the main beneficiary group and identify an adequate stream of funding.

The role of development co-operation in supporting the scaling up of ocean-relevant risk management tools

In the examples above, development partners provided critical support that allowed innovative products for managing increasing ocean risks to be tested and developed and therefore to generate funding for the conservation of marine ecosystems that contribute to reducing ocean risks. It is positive that development partners such as Canada and some multilateral institutions have been at the forefront of these experiments, but a larger number of champions is required. Critically, development partners have a role to play in helping bring together different communities such as governments, conservation experts, the insurance industry and investors to help develop a common understanding of the challenges at stake and innovative solutions that can align interests. Technical assistance, support for building data systems, and capacity building to develop expertise for managing these new instruments will be essential.

Investment funds and products for the ocean

Current market

In the context of sustainable ocean economies, making investments sustainable implies a dual challenge: scaling up investment in the sustainable use and conservation of the ocean and scaling down/ phasing out investment in unsustainable economic activities that accelerate irreversible destruction of marine ecosystems. For decades already, investments in industrial overfishing, dumping of chemical and plastic waste into the ocean, and other damaging activities have pushed the ocean to unprecedented conditions. The expected expansion of the global ocean economy could further aggravate the situation. The global ocean economy is expected to grow faster than the rest of the world economy (OECD, 2016[33]) and a race to secure extraction rights – including for oil, gas and seabed mining – has already started. Growth in ocean investment is by no means destined to be sustainable in environmental and social terms. The characteristics of several ocean sectors suggest that harmful impacts are actually more likely (Jouffray et al., 2020[34]).

On the other hand, fostering sustainable ocean economies presents tremendous business and investment opportunities that range from climate resilient infrastructure to innovative businesses addressing plastic pollution, sustainable tourism, sustainable fisheries and decarbonised shipping, to mention only a few. Sustainability is not earned at the expense of profitability. The World Bank (2017[35]) estimates that the global fisheries industry would earn an extra USD 83 billion every year if fisheries were managed sustainably. But there is a need to rethink short-term profit (or return on investment) as the key metrics for business and investment success.

While it is difficult to estimate how much investment goes into ocean-based industries and ecosystems, and how much of that is sustainable, a few dedicated impact investment funds have emerged in the last five years that are directly relevant for attracting commercial capital for sustainable ocean economies. Investments currently focus mainly on marine conservation, sustainable fisheries, and marine plastic litter and/or circular economy. These funds been developed under the impetus of impact investors with support from philanthropies and development partners. The United States, for instance, has provided loan guarantees to several of these funds, including the Althelia Sustainable Ocean Fund, the Meloy Fund and Circulate Capital's Ocean Fund. These new impact investment funds are briefly described in Box 5.1.

Box 5.1. Impact investing for sustainable ocean economies

The Althelia Sustainable Ocean Fund is a public-private partnership established by Althelia Ecosphere, with the support of Conservation International and with technical and scientific advice from the Environmental Défense Fund. It aims to invest in companies able to deliver marine conservation, improved livelihoods as well as attractive economic returns. The Fund's portfolio will focus on circular economy, sustainable seafood and ocean conservation. Key targeted areas for investments include coastal fisheries, sustainable aquaculture, the seafood supply chain and other select coastal projects. Target impacts include improved food and climate security, livelihoods and ecological biodiversity. Through blended financing models, the Fund seeks to crowd in sustainable ocean related investments, by reducing their risks and improving returns. The Fund's requirement of sustainable covenants and the subscription of contracts will ensure the compliance of the environmental and social safeguards. In 2019, the fund's achieved its first closing of USD 100 million of capital, with commitments from leading institutional investors, including the European Investment Bank and the Inter-American Development Bank, among others. The United States Agency for International Development (USAID) participated providing a USD 50 m guarantee facility to cover eligible projects in the portfolio. In 2016 and 2017, the Fund identified potential investments in Belize, Bangladesh, Colombia, Chile, Ecuador, Honduras, Mexico and Madagascar (Althelia Ecosphere, 2020[36]).

The Meloy Fund for Sustainable Community Fisheries is a USD 20 million impact investment fund, of which USD 6 million was committed on concessional terms by the GEF. The Fund was established to invest through debt or equity in fishing-related enterprises that support the recovery of coastal fisheries in Indonesia and Philippines. Interventions target supply chain and production inefficiencies, waste reduction, aggregation, and value added processing. The Fund also invests in projects to offset fishing pressure on fish populations, such as ocean-based aquaculture. This allows for stock recovery and helps fishers access alternative, complementary income sources. The average investment amounts to USD 500 000 to USD 2 million with a duration of five to seven years. In 2016, Rare Conservation, of which the Fund is a subsidiary, closed a USD 1-million, five-year investment in Meliomar, a Philippines-based seafood company (Duke University/Environmental Defense Fund, 2018[37]).

> **Encourage Capital** is a social investment firm that aims to recover the economic value of fisheries that have been affected by poor management policies, restore ocean ecosystems, and increase the sustainable seafood supply in the pilot countries of Brazil, Chile and Philippines. It was established with support from the Rockefeller Foundation and Bloomberg Foundation. With USD 10 million in support from Zoma Capital, the family investment office of Ben and Lucy Ana Walton, Encourage capital established Pescador, a sustainable seafood investment holding company. The company's focus is the recovery of hake fishery in Chile, which collapsed in the early 2000s (Inamdar et al., 2016[38]).
>
> USAID and the Circulate Capital impact investment firm launched an initiative to attract private capital to combat plastic pollution in the ocean across the Indo-Pacific region. USAID provided a USD 35-million 50% loan portfolio guarantee to Circulate Capital to mobilise private capital by lowering the risk of loss and making investment more appealing. The initiative aims to be a powerful instrument for developing a market around waste management, recycling and the circular economy in the Indo-Pacific region by financing companies, innovation and infrastructure projects to improve waste management systems. At least half of the total investment will be allocated to Indonesia, Philippines, Viet Nam and Sri Lanka, four of the countries most impacted by ocean plastic. To date, the fund created through this initiative has attracted USD 100 million in commitments from some of the world's leading businesses.
>
> With the aim to promote impact investing and blended finance solutions for conservation of marine biodiversity, Blue Finance is developing a suite of investments in the Caribbean, including Antigua and Barbuda, the Dominican Republic, and Saint Kitts and Nevis. Blue Finance operates under the umbrella of the UN in partnerships with government, investors and other stakeholders to ensure sustainable financing and efficient management of marine protected areas (MPAs). The Blue finance initiative expects to restore marine and terrestrial biodiversity, improve tourism attractiveness, and create significant job opportunities in the tourism, fishery and agro-forestry sectors. Blue Finance expects to restore marine and terrestrial biodiversity, improve tourism attractiveness, and create significant job opportunities in the tourism, fishery and agro-forestry sectors. It also promotes public-private partnerships for the management of MPAs, whereby the private sector is expected to provide the majority of required funds to improve and manage an area in exchange for a return on investment, mainly through fees and innovative tourism products. In this scheme, national governments maintain their core functions, with responsibilities over the regulation and enforcement of uses and zoning, the establishment user fees and the maintenance of specific onshore facilities. Funds are used mainly to finance the upfront capital expenditures and innovative tourism activities. (Pascal et al., 2018[39]).

Opportunities and challenges for scaling up impact investing for sustainable ocean economies

While still at an early stage, new impact investing funds focused on the ocean are helping fill the financing gap for marine conservation and sustainable ocean-based economic activities and demonstrating the feasibility of achieving environmental, social and financial returns at the same time.

Scaling up impact investing for sustainable ocean economies faces several challenges common to the broader impact investing industry and some specific challenges. Common challenges include the need to develop data, definitions and standards, the ticket size and thus the need to structure and aggregate projects so that they can accommodate demand from larger investors, and the need to familiarise investors with new products. These challenges may be exacerbated by the fact that, while impact investment is in general fairly recent, impact investing in sustainable ocean economy activities is even more recent, although growing rapidly. As discussed, there are currently no recognised standards and classifications for what should be defined as sustainable across ocean-based economic activities. Therefore, a critical challenge becomes the identification of a pipeline of bankable projects with a clear and proven sustainability profile.

Impact investing in marine conservation may face specific challenges relating to the identification of revenue streams able to generate returns that meet the appetite of investors. Overall, private investment in marine biodiversity and ecosystem services is in an early stage of development and practical experience is very limited. While there are numerous examples of impact investments in terrestrial environments, such as those focusing on watershed and forest ecosystem services, initiatives focusing on marine ecosystems are rare and not well documented. The lack of a track record of solid investment opportunities is thus a challenge to scaling up investments (Pascal et al., 2018[39]).

Suggestions for re-orienting investments towards sustainability

The global impact investing market is growing fast and reached an estimated USD 502 billion in 2019 (Global Impact Investing Network, 2019[40]). Yet, this is still a tiny fraction of total global investments. The blue bonds market is in its infancy. The green bond market, that some point to as the reference and target market for blue bonds, amounts to less than 0.6% of the total bond market globally despite sustained growth in the past decade.

Therefore, a two-track approach is needed. On one hand, it is necessary to increase the contribution of these niche investments to sustainable ocean economies and, as discussed so far in this chapter, to explore how new and existing financial mechanisms can grow finance for sustainable ocean-related economic activities and ocean conservation. Table 5.2 summarises the range of new financial instruments discussed above and how development co-operation can help replicate them and scale them up. On the other hand, it is necessary to mainstream sustainability concerns to the bulk of global investments and corporate finance in ocean-based industries and those that have either direct or indirect impacts on the ocean. This is particularly important because there is growing appetite from investors to invest in the ocean and yet most investors are not aware of their investments' effects on the marine environment (Fritsch, 2020[9]). Investors are also not aware of how degrading ocean ecosystems may subsequently affect their portfolios' performance and value. In other words, the financial system is building up liabilities of which it is not aware.

Table 5.2. How development co-operation providers can help replicate and scale up innovative financial mechanisms and scale down unsustainable investments

Blue bonds	- Support the design of a clear mission-oriented strategy for the sustainable ocean economy - Support the design of a financing plan that the blue bonds will be part of, in conjunction to other financing instruments. - Support the development of a pipeline of projects responding to the mission-oriented strategy - Structure concessional finance packages (such as grants, guarantees, etc.) as a mean of credit enhancement to make borrowing affordable for the country - Enhance domestic capacities to manage complex borrowing options and meet the costs of bonds design
Debt-for-ocean swaps	- Provide expertise for structuring debt conversions - Engage early funding commitments to attract political and investor interest - Help obtain creditor consent to conduct the debt conversion
Blue carbon schemes	- Incorporate livelihoods aspects in the design of blue carbon projects - Conceptualise future projects based on ecosystems other than mangroves, such as marshes and seagrass - Include carbon stored within the soil of blue carbon ecosystems in the project's potential
Ocean relevant risk management tools	- Build data systems and develop expertise for managing these new instruments will be essential - Develop a common understanding among different communities of the challenges posed by ocean risks - Explore and develop innovative products to manage increasing ocean risks and to fund the conservation of marine ecosystems that reduce ocean risks
Scaling up sustainable ocean investments	- Improve the investment size and aggregation of ocean projects - Develop a pipeline of bankable projects that respond to clear and verifiable sustainability criteria and are aligned to partner countries' priorities

Scaling down unsustainable ocean investments	- Integrate ocean sustainability requirements in all ODA lending and all development finance institutions (DFIs) lending - Support the adoption of the Sustainable Blue Economy Finance Principles and the integration of ocean-sustainability requirements by international financial institutions - Advocate for the adoption of the integration of ocean sustainability requirements by exchange listing and other financial market regulations

Source: Authors.

Requiring integrated reporting of both financial and non-financial information would enable investors, financiers and other stakeholders to better assess firm performance and risks (Jouffray et al., 2019[41]). Financial reporting would need to include the assessment and management of ocean risks, including both the impact of investments on ocean and the risks on investments from degrading ocean health. Non-financial reporting would need to enhance disclosure on practices. For instance, for seafood companies, given the widespread prominence of IUU fishing, with one in every five fish sold being stolen (FAO, 2017[42]) and equivalent to USD 23 billion each year, introducing requirements on origin and traceability as well as on human and labour rights will be critical.

A number of principles for responsible and sustainable investments have been developed recently that could help refocus investments. These are, however, voluntary commitments that call upon the good will of companies and investors. To be effective, they will need to be subscribed to more broadly and linked to clear and solid implementation and monitoring frameworks. Relevant principles include:

- The Sustainable Blue Economy Finance Principles, developed in 2017 by a partnership of the European Commission, World Wildlife Federation, the Prince of Wales's International Sustainability Unit in the United Kingdom and the European Investment Bank.
- The Principles for Investment in Sustainable Wild-Caught Fisheries, launched at the World Ocean Summit 2018.

Finally, identifying leverage points in the financial system will be important to redirect investments towards more sustainable practices, within and across industries. In their recent study, Jouffray et al. (2019[41]) identify bank loan covenants, stock exchange listing rules and shareholder activism as potential financial sector leverage points for increasing sustainability in the seafood industry. While developed with a specific reference to the seafood industry, these levers could be relevant to other industries that produce impacts on ocean. Given that medium and larger companies that are not listed on exchanges rely heavily on bank loans, sustainability requirements could be a particularly effective instrument through which banks might encourage companies to refrain from harmful practices or to adopt sustainable behaviour. An example is the Louis Dreyfus Company, which agreed to a USD 750-million loan with an interest rate that is linked to the company's sustainability performance; the higher the sustainability performance rating, the more the interest rate goes down (Jouffray et al., 2019[41]).

Introducing sustainability criteria in exchange listing rules could be particularly effective, as just two of the world's ten largest stock exchanges now require some environmental and social reporting as a listing rule companies (Sustainable Stock Exchanges Initiative, 2019[43]). Shareholder activism also could be leveraged to enhance the weight of sustainability considerations in corporate decisions, especially in light of documented increasing levels of activism among shareholders (Jouffray et al., 2019[41]).

The role of development co-operation in helping re-orient private investments

A few development partners have supported the establishment and implementation of new impact investing funds and products to channel commercial finance towards marine conservation and sustainable ocean-based activities. While such support has come mainly from development partners that have more flexibility in their legislation and policies to use financial instruments other than standard ODA grants and loans, such standard grants and loans can also be used catalytically to crowd in private investments. Moreover, development partners can do much to help developing countries tackle issues around investment size and the aggregation of projects and to develop a pipeline of bankable projects that respond to clear and verifiable sustainability criteria and are aligned to partner countries' priorities. Technical assistance and support for the preparation costs could be provided to develop a pipeline of investment opportunities with an attractive investment size. This is a critical issue because investment tickets, especially for marine conservation, are usually smaller than institutional investors' minimum investment size but larger than many individual impact investors' desired allocation (Pascal et al., 2018[39]). Innovative deal and fund structures consolidating projects could help accommodate investors' needs and facilitate the matchmaking.

Development partners have a further role to play in integrating sustainability requirements into traditional financial services and investments, whether in financial markets (e.g. stocks and bonds) or in credit markets (e.g. loans or bonds). To help redirect and re-orient private finance towards sustainable ocean-related businesses and activities and support the co-creation of new sustainable markets and products, development co-operation providers need to use ODA catalytically to enhance the risk-return profiles of sustainable investments and supporting the development of new financial instruments. They should also use their financial levers, such as grants and loans requirements, to push businesses and actors to adopt sustainable practices and standards. Development partners have an additional role in supporting financial system reforms and regulations to phase out unsustainable investments, including through sustainability requirements in stock exchange listing rules. Specific suggestions include:

- use ODA catalytically to improve the commercial viability of investments in sustainable activities and businesses, helping to create new sustainable products and markets including through new investment vehicles and instruments
- integrate ocean sustainability requirements in all ODA lending and in all development finance institution (DFIs) lending (not all of which is concessional in ODA terms)
- support the adoption of the Sustainable Blue Economy Finance Principles and the integration of ocean sustainability requirements by international finance institutions, which bilateral development partners can influence as they are members and shareholders of these institutions
- advocate for the adoption of the integration of ocean sustainability requirements in exchange listing rules and other financial market regulations to refocus investments to ocean-based industries towards sustainability
- strengthen independent assessments of the impacts of financial flows to the ocean economy such as through international and research institutions.

References

Althelia Ecosphere (2020), *Sustainable Ocean Fund (webpage)*, https://althelia.com/sustainable-ocean-fund/. [36]

Asiedu-Akrofi, D. (1991), "Debt-for-Nature Swaps: Extending the Frontiers of Innovative Financing in Support of the Global Environment", *The International Lawyer*, Vol. 25/3, https://scholar.smu.edu/cgi/viewcontent.cgi?article=2820&context=til. [12]

Caribbean Development Bank (2018), *Financing the Blue Economy: A Caribbean Development Opportunity*, http://www.caribbeanhotelandtourism.com/wp-content/uploads/2018/09/CDB-Financing-the-Blue-Economy-A-Caribbean-Development-Opportunity-2018.pdf. [10]

Convergence (2017), *Seychelles Debt Conversion for Marine Conservation and Climate Adaptation Case Study*, https://www.convergence.finance/resource/3p1S3pSTVKQYYC2ecwaeiK/view. [14]

Duke University/Environmental Defense Fund (2018), *Financing Fisheries Reforms: Blended Capital Approaches in Support of Sustainable Wild-Capture Fisheries*, https://nicholasinstitute.duke.edu/publications/financing-fisheries-reform-blended-capital-approaches-support-sustainable-wild-capture. [37]

FAO (2017), "Seafood traceability for fisheries compliance: Country-level support for catch documentation schemes", *FAO Fisheries and Aquaculture Technical Paper*, No. 619, Food and Agriculture Organization, Rome, http://www.fao.org/3/a-i8183e.pdf. [42]

Fourqurean, J. et al. (2012), "Seagrass ecosystems as a globally significant carbon stock", *Nature Geoscience*, Vol. 5, pp. 505-509, https://doi.org/10.1038/ngeo1477. [19]

Fritsch, D. (2020), *Investors and the Blue Economy*, Credit Suisse, London, https://www.esg-data.com/blue-economy. [9]

Fuller, F. et al. (2018), *Debt for Climate Swaps: Caribbean Outlook*, Climate Analytics, Berlin, http://www.undp.org/content/sdfinance/en/home/glossary.html. [16]

Global Impact Investing Network (2019), *Sizing the Impact Investing Market*, https://thegiin.org/assets/Sizing%20the%20Impact%20Investing%20Market_webfile.pdf. [40]

Herr, D. et al. (2018), "Coastal blue carbon and Article 6: Implications and opportunities", Climate Focus, Amsterdam, https://climatefocus.com/sites/default/files/20181203_Article%206%20and%20Coastal%20Blue%20Carbon.pdf. [25]

Herr, D., E. Pidgeon and D. Laffoley (2012), *Blue Carbon Policy Framework 2.0*, International Union for Conservation of Nature, Gland, Switzerland, https://portals.iucn.org/library/sites/library/files/documents/2012-016.pdf. [26]

IMF (2016), *Small States' Resilience to Natural Disasters and Climate Change – Role for the IMF*, International Monetary Fund, Washington, DC, https://www.imf.org/en/Publications/Policy-Papers/Issues/2016/12/31/Small-States-Resilience-to-Natural-Disasters-and-Climate-Change-Role-for-the-IMF-PP5079. [18]

Inamdar, N. et al. (2016), *Developing Impact Investment Opportunities for Return-Seeking Capital in Sustainable Marine Capture Fisheries*, World Bank, Washington, DC, http://www.wildernessmarkets.com/wp-content/uploads/Fish-Finance-Paper-final-clean.pdf. [38]

Indonesia Ministry of Finance (2019), *Green Sukuk Issuance, Allocation and Impact Report*, https://www.sdgphilanthropy.org/system/files/2019-02/Green%20Suku%20Issuance%20-%20Allocation%20and%20Impact%20Report%20.pdf. [8]

Jouffray, J. et al. (2020), "The Blue Accelleration: The Trajectory of Human Expansion into the Ocean", *One Earth*, Vol. 2/1, pp. 43-54, https://doi.org/10.1016/j.oneear.2019.12.016. [34]

Jouffray, J. et al. (2019), "Leverage points in the financial sector for seafood sustainability", *Science Advances*, Vol. 5/10, http://dx.doi.org/DOI: 10.1126/sciadv.aax3324. [41]

Kattel, R. et al. (2018), "The economics of change: Policy and appraisal for missions, market sharping and public purpose", *Working Paper Series*, No. 2018-06, UCL Institute for Innovation and Public Purpose, London, https://www.ucl.ac.uk/bartlett/public-purpose/sites/public-purpose/files/iipp-wp-2018-06.pdf. [1]

Morgan Stanley Institute for Sustainable Investing (2019), *Blue Bonds: The Next Wave of Sustainable Bonds*, https://www.morganstanley.com/content/dam/msdotcom/ideas/blue-bonds/2583076-FINAL-MS_GSF_Blue_Bonds.pdf. [7]

Murray, B. et al. (2010), "Payments for blue carbon: Potential for protecting threatened coastal habitats", *Policy Brief*, No. 10-05, Nicholas Institute for Environmental Policy Solutions, Duke University, Durham, NC, https://nicholasinstitute.duke.edu/sites/default/files/publications/blue-carbon-paper.pdf. [27]

Murray, B. et al. (2011), *Green Payments for Blue Carbon: Economic Incentives for Protecting Threatened Coastal Habitats*, Nicholas Institute for Environmental Policy Solutions, Duke University, Durham, NC, https://nicholasinstitute.duke.edu/sites/default/files/publications/blue-carbon-report-paper.pdf. [20]

OECD (2018), *Making Blended Finance Work for the Sustainable Development Goals*, OECD Publishing, Paris, https://dx.doi.org/10.1787/9789264288768-en. [2]

OECD (2016), *The Ocean Economy in 2030*, OECD Publishing, Paris, https://dx.doi.org/10.1787/9789264251724-en. [33]

OECD DAC (2020), *Amounts mobilised from the private sector for development*, https://www.oecd.org/dac/financing-sustainable-development/development-finance-standards/mobilisation.htm (accessed on 19 June 2020). [3]

Pascal, N. et al. (2018), "Impact Investment in Marine Conservation", *Duke Environment Law & Policy Forum*, Vol. XXVIII/199, https://pdfs.semanticscholar.org/6d2b/e4faf94728ee781378054889fb4a41142eb3.pdf. [39]

Pörtner, H. et al. (eds.) (2019), *Summary for policymakers*, Intergovernmental Panel on Climate Change (IPCC), Geneva. [21]

Reguero, B. et al. (2019), "The risk reduction benefits of the Mesoamerican Reef in Mexico", *Frontiers in Earth Science*, Vol. 7, http://dx.doi.org/10.3389/feart.2019.00125. [32]

Roth, N., T. Thiele and M. von Unger (2019), *Blue Bonds: Financing Resilience of Coastal Ecosystems*, 4Climate, Bereldange, Luxembourg, https://www.4climate.com/dev/wp-content/uploads/2019/04/Blue-Bonds_final.pdf. [5]

Secretaria de Ecologia y Medio Ambiente Quintana Roo (2019), *Seguro paramétrico para arrecifes y playas en Quintana Roo*, Secretaría de Ecología y Medio Ambiente Quintana Roo, https://qroo.gob.mx/sema/seguro-parametrico-para-arrecifes-y-playas-en-el-estado-de-quintana-roo. [30]

Silver, J. and L. Campbell (2018), "Conservation, development and the blue frontier: the Republic of Seychelles' Debt Restructuring for Marine Conservation and Climate Adaptation Program", *International Social Science Journal*, https://doi.org/10.1111/issj.12156. [13]

Spalding, M. et al. (2014), "The role of ecosystems in coastal protection: Adapting to climate change and coastal hazards", *Ocean & Coastal Management*, Vol. 90, pp. 50-57, http://dx.doi.org/10.1016/j.ocecoaman.2013.09.007. [28]

Sustainable Stock Exchanges Initiative (2019), *Stock Exchange Database*, https://sseinitiative.org/data/. [43]

The Nature Conservancy (2020), *Naturevest - Debt Conversions for Marine Conservation and Climate Adaptation*, https://www.nature.org/en-us/about-us/who-we-are/how-we-work/finance-investing/naturevest/ocean-protection/. [15]

The Nature Conservancy (2020), *Playbook for Climate Action*, https://www.nature.org/en-us/what-we-do/our-insights/perspectives/playbook-for-climate-action/. [31]

Thrush, S. (ed.) (2012), "Estimating global 'blue carbon' emissions from conversion and degradation of vegetated coastal ecosystems", *PLoS ONE*, Vol. 7/9, p. e43542, http://dx.doi.org/10.1371/journal.pone.0043542. [22]

UN (2016), *Proposal on Debt for Climate Adaptation Swaps: A Strategy for Growth and Economic Transformation of Caribbean Economies*, United Nations Economic Commission for Latin America and the Caribbean, Trinidad and Tobago, https://repositorio.cepal.org/bitstream/handle/11362/40253/LCCARL492_en.pdf?sequence=1&isAllowed=y. [17]

UN (2015), *United Nations Office for DIsaster Risk Reduction - Direct and indirect losses (webpage)*, https://www.preventionweb.net/risk/direct-indirect-losses. [29]

United Nations Global Compact (2020), *Blue Bonds: Reference Paper for Sustainable Ocean Investments*, https://ungc-communications-assets.s3.amazonaws.com/docs/publications/Blue-Bonds-Reference-Paper-for-Sustainable-Ocean-Investments.pdf. [6]

Vanderklift, A. et al. (2019), "Constraints and opportunities for market-based finance for the restoration and protection of blue carbon ecosystems", *Marine Policy* 107, https://doi.org/10.1016/j.marpol.2019.02.001. [23]

Walsh, M. (2018), *Ocean Finance: Definition and Actions*, https://gallery.mailchimp.com/b37d1411f778c043250da5ab5/files/f1a910e2-32f9-4aed-ad35-e2bccab6cf12/Ocean_Finance_Definition_Paper_Walsh_June_2018.pdf. [11]

World Bank (2018), "Sovereign Blue Bond Issuance: Frequently Asked Questions", https://www.worldbank.org/en/news/feature/2018/10/29/sovereign-blue-bond-issuance-frequently-asked-questions. [4]

World Bank (2017), *The Sunken Billions Revisited: Progress and Challenges in Global Marine Fisheries*, World Bank, Washington, DC, http://dx.doi.org/10.1596/978-1-4648-0919-4. [35]

Wylie, L., A. Sutton-Grier and A. Moore (2016), "Keys to successful blue carbon projects: Lessons learned from global case studies", *Marine Policy*, Vol. 65, pp. 76-84, http://dx.doi.org/10.1016/j.marpol.2015.12.020. [24]

Notes

[1] The information on private finance mobilised for the sustainable use of the ocean was retrieved from the 2016 OECD-DAC Survey on amounts mobilised from the private sector in 2013-17 by official development finance interventions. Based on the information registered in the database, projects selected were those classified under purpose codes related to ocean activities or ocean sustainability (sanitation, water supply, fishery development, fishery policy, flood prevention, river basins development, waste management, water transport, water resources conservation) and those for which the title included one of the ocean keywords prioritised (ocean, water, aquaculture).

[2] The impact investors are Nuveen, the asset management arm of TIAA (which will include the bond in the TIAA-CREF Social Choice Bond Fund), Prudential Financial and Calvert Impact Capital.

[3] The Paris Club is a group of officials from 22 major creditor countries who negotiate co-ordinated solutions to the payment difficulties experienced by debtor countries.

www.ingramcontent.com/pod-product-compliance
Lightning Source LLC
LaVergne TN
LVHW061935070526
838199LV00060B/3837